THE WARRIOR FROM SIATA

A JOURNEY THROUGH TIME

THE THIRD BOOK IN THE
WARRIOR FROM SIATA SERIES

By
Ken Bastian

Table of Contents

Chapter 1

A family is a multi-faceted, complex organization. It's a small world created by love, made up of individuals with disparate ages, backgrounds, ideologies, and personalities. It's supposedly managed by the adults, the older and allegedly wiser members of the organization who are supposed to know what's best for everyone. At least, that's what the book says. The truth is, it isn't a democracy, nor is it a dictatorship. Most of the time, nobody knows what's best for anyone. It can be, and too often is, even in the best of families, an organization that at times exists in chaos and defies any problem-solving solutions.

All families are unique. Mine? I think it's safe to say that mine is a little more unique than most. My husband, Apu, came from twenty-two thousand years ago, born on a continent that no longer exists. We met and fell in love there—and then. We have a daughter, conceived and born here and now. I named her after a special friend and warrior—another one from twenty-two thousand years ago, who is rightfully back in her own time. My adopted daughter, Masana, came from twelve thousand years ago, and her culture is gone, lost in a tragedy that still haunts my dreams to this day.

That blend of times and cultures is sure to create chaos, but things have actually been relatively peaceful in our family. Apu and Masana had adapted to my culture quite well, with a few quirks here and there. We'd come up with a pretty good backstory for who they are and where they came from, and even came up with fake identification for them so they could function in our society. And they did. Some things, however, like human emotions, are eternal and cross all lines of time and culture.

It started out as a really great day but didn't stay that way. I had some exciting news to share when I got home from work, but no one greeted me. Sina was in her room with the door shut, sobbing out loud. Masana was doing her best to comfort her, as any sister would. Apu was fixing dinner but paused long enough to give me a kiss and tell me Sina was heartbroken. The guy she had a crush on had put her down in a big way. And he didn't even try being nice about it. At least, that's what I was told.

And the most devastating part is that I guess I'm to blame.

She's fifteen, old enough to think about boys, and this one was, as she put it, "to die for." I'd never met the kid, knew of him only through what she'd told me, so how was I to blame when it didn't work out? I had a pretty good idea, so I went to her room to talk, but she screamed for me to get out. Masana gave me a warning look that said this wasn't the right time, so I went back downstairs to escape.

That put the celebration I'd planned for the evening completely out of mind.

Apu was busy with food preparations, trying to ignore the sobbing that echoed down the stairs. He doesn't like me to help with the cooking for the most part, so I sat in the living room, hurt over Sina's despair. Sina was a great kid. She'd never been rude to me before, so this came as a complete shock. Apu forced a grin and told me not to worry, that she'd come around after she cried it out.

"But what did I do?" I asked. But I knew the answer. I asked him in Samoan because he was still struggling with English. He knew enough to go shopping or run errands, but that was enough for him. Most talk at home was in his language, and he didn't go out without me very often.

"It is the same thing again," he replied, then went back to fixing dinner.

Masana had quickly picked up English and adopted the new culture she found herself in. She and Sina had become best friends despite their age difference. Masana was now twenty-four, a beautiful young woman who had found love herself, so we were trying to make the arrangements for a wedding. Her ID was fake, since there was no way to get official ones, but we hoped it would work. Apu's ID was also fake, and for the same reason.

We were a strange family. Sina and I were normal members of society and could get official IDs, but Apu and Masana didn't legally exist. As far as the official record was concerned, they'd never been born. No matter how extensive your background check is, you won't find a thing on either of them. Even with a fake ID, I couldn't add them to my company insurance plan, so any medical expenses were strictly on a cash basis. Fortunately, neither of them had required anything serious. That was going to be a problem for her husband when Masana got married, so she'd told him the truth about her, and he was too much in love to care.

Masana and Sina eventually came downstairs when dinner was ready. Sina's eyes were red and puffy, but she'd apparently cried herself out, and her sorrow gave way to hunger. She came straight to me, gave me a fierce hug, and looked me in the eye. "I'm sorry for being mean to you, Mom. It's just that he made me so mad."

"I'm guessing it had something to do with my trips into the past. Am I right?"

She nodded. "He thinks you're crazy to make claims like that, and he doesn't want anything to do with the daughter of a crazy woman."

"And how do you feel? Do you think I'm crazy?"

"Of course not," she said, looking at me as if I was crazy to even think such a thing. "My dad's from the past, and Masana, my sister and best friend. They're both from the past, and you brought them here. No, Mom, you've gone back, and each time you've brought back someone special. Next time maybe you could bring me a boyfriend? I'm not having much luck with that."

"Don't worry, honey. You've got lots of time for that. There'll be someone special for you when the time's right. I'm sorry I screwed this one up for you, though it sounds to me like he's a total jerk. And a loser, since he missed his chance with a great girl like you. You don't consider yourself the daughter of a crazy woman, then?"

"He's the one who's screwed up," she chuckled, color coming back to her pale cheeks. "Who made him a judge, anyway? It must make him feel big to put people down for something he knows nothing about."

"Some people are that way," I agreed, "but give him a break. After all, time travel is thought to be impossible. At the very least, it's way out there. You know, in la-la land."

"I suppose so," she said, filling her plate with about half the mashed potatoes Apu had fixed, then putting some back when she realized she hadn't left enough for the rest of us. She likes mashed potatoes dripping with butter or brown gravy. "It must be cool traveling through time, though."

"There's really not much to it. One second you're here, the next somewhere else in time and space, so there's really no 'travel' involved. In an instant, you're facing a different culture and people. You don't understand them, and they don't understand you, so it's a matter of adapting. For the most part, people are pretty much the same no matter when or where they come from. They just speak different languages. They want love, peace, shelter, food, freedom, and security, just like us."

"Yeah, but it's still gotta be cool to live in another culture like the ones you've been to. I envy you."

I laughed. "I wouldn't trade those experiences for anything in the world, but I'm glad it's over and I won't have to do it again." That was stated with a serious air of finality, and I meant it. "Now, before we start dinner, I have an announcement to make. Raise your glasses in celebration, because this morning I was informed that I am now a tenured professor at Boise State University."

Sina understood the meaning of that more than anyone else in the family and shrieked with joy. "You're tenured, Mom? That's cool. Congratulations. That blue football field will be bluer than ever with you in it."

"Thanks," I said, blushing from pleasure rather than shyness, "but my position has nothing to do with football." It was a big achievement for me, the professor who couldn't get a job for years until Boise gave me a chance eight years ago. I had baggage in the form of rumors floating around about claims of time travel, a fraudulent claim as far as everyone was concerned, including Sina's "boy to die for." Since then, I'd taught full time, helped to establish a monument to the local Native Americans, and successfully handled four digs of pre-Columbian sites.

I had a higher than average rate of students graduating in my classes, an excellent evaluation and review from both students, peers, and the dean. I'd managed to publish nine papers on early American civilizations, carefully avoiding any mention of prehistoric Polynesia, which would have brought back the whole time travel issue, although that was the passion I truly wanted to pursue. I'd worked hard and spent a lot of money and man-hours to get recognized, and it had finally paid off. My future was more secure as a result of being tenured, and with the crazy family I had, security was a definite plus.

Thinking of the future always made me think about the past. My first experience in the ancient past, although it had been an accident caused by those simple-minded Little People who can't keep track of their unusual devices, had been difficult but rewarding beyond belief. The second one even more difficult but not rewarding in the slightest, except that I got Masana out of it. It had ended with the complete destruction of the place and people I loved deeply. I'd always thought of the biblical Flood as a matter of rising rainwater until I witnessed it firsthand and saw devastation far more brutal than backed up sewers.

That one left a bitter taste in my mouth, hoping beyond hope that I'd never hear from the Little People again, never take another journey into the past. It was too painful, caused too much emotional hurt to think about going back. I couldn't do it again. I wouldn't do it again. Never!

But then, on occasion, I'd take out my spear, the tupa, run my hands gently over its surface and trace the squiggles on the blade. And I'd remember those cherished days I'd spent with Sina. Sina and that spear had changed my life, had made me a better person. The People of Tufutafoe had given me a completely different outlook about myself and what I was capable of. I knew deep inside me, at a personal level I tried to conceal regardless of how deep it went, that I wanted to feel that way again.

In my world, there's no need for a female warrior, especially one with a wooden spear as her weapon. Yes, I'd killed a man with a gun during my last trip to the past, but that was luck. The gun I'd used was unlike any I'd ever seen in modern times, and I wasn't unfamiliar with guns, though no one would call me a marksman. No, my skills had nothing to do with guns.

I was skilled with the spear, with which I'd fought battles and killed enemies, both human and beast. I'd been trained in face-to-face combat, not shooting at distant targets. I had no desire to kill again, but the desire to take that spear once again into action was a burning itch. Not against a saber-tooth tiger, though. Definitely not that again. But to spar with someone, to practice with that special weapon again, was like a decadent dream.

There was no getting around it. That spear, which to anyone else would be nothing more than an elaborate decoration, was an important part of my life. An unmistakable part of my psyche was centered around it. It, and Sina, had given me the courage and strength to stand tall amid all the challenges that came my way as a result of my journeys into the past. The first one, especially. The second one had been kept more secretive despite Masana coming back with me. Very few people knew about it, and I wanted to keep it that way. It concerned me that Masana had told her fiancé the truth about her life, where and when she came from. Did he actually believe her, and how far would it spread from there?

The courage and self-esteem I'd gained in those excursions helped me in the present, too. I had a completely different outlook, not only about myself but about life in general and everyone I came in contact with. That spear and my relationship with Sina had made me a more complete, happy person no matter what obstacles came my way.

Despite the urges to use the spear again, I hated even thinking about another encounter with ancient cultures and challenges. The past could be exciting and exhilarating, but it could also be dangerous and loathsome. It could bring love and hate, joy and sorrow, all at the same time. Blessing and curse. That was the past I knew, so I cherished the blessing part and feared with all my heart the curse that came with it.

Apu, the man I loved and the father of Sina, was one of those blessings. So was Masana, who I thought of as my own daughter. But seeing an entire civilization, a huge continent, and most of the world wiped out in a conflagration like nothing the average mind could even conceive of was the worst of the curses. Having to kill beasts and enemy warriors was unpleasant, but that paled in comparison to the destruction I'd witnessed on that last trip back.

I was happy, therefore, to live in a time and place where wars were fought far away by professional warriors with sophisticated weapons rather than up front and personal the way I'd had to fight. They didn't have to muck their way through blood and gore the way we did. At least, I hoped they didn't, but war is war and anything can happen.

I was still trying to erase the memories of bloody battles, to forget that I'd had to kill and watch people who were special to me die. That wasn't the pleasant part of the past, but there had been some good times as well as the pain and death. The love I found in Apu, the love I felt for the original Sina, and the love I felt for my adopted daughter, Masana, were all presents from the past. And in the final analysis, they more than made up for the negatives.

Still, I wasn't going back again, come hell or high water. There was nothing that could make me risk the heartache and pain I'd found in the past despite the special loves that came my way. No, the past was going to stay in the past, no matter what. I was a girl of this time and place, and intended to stay where I was supposed to be.

Chapter 2

Boise doesn't get a lot of snow. Most snow storms consist of a few inches at most, and that too dry to pack into a decent snowman. The snow that fell this Saturday wasn't like that at all. We got four to five inches, and it was wet enough to be workable. Masana, who never got tired of playing in the snow when we got it, was out building a snowman with Sina. She was having a blast, even though she had to come in periodically to warm up. They both had good coats, insulated underwear, and gloves, but the cold still got through after a while, so they took frequent breaks. Their snowman, bigger than any they'd built before, was coming along nicely.

Apu and I were putting away the Christmas decorations. It wasn't a big job since I don't do a lot of decorating, but Apu thought the tree was exciting with all that stuff on it. For him, the holidays were amazing, so I tried not to dampen his spirits. My attitude about Christmas and New Year's was lackluster at best, but I had to act excited for Apu and the kids.

Actually, they weren't kids anymore, a fact I was still trying to come to grips with. Despite their age difference, they had become best friends, and Sina looked to Masana as her role model. Both had grown to be lovely young women, and I had a deep sense of pride in them. They had the grace of a woman and the enthusiastic excitement of a child. It was an honor to be their mom.

I was watching them work on the snowman through the front window when it happened. Masana stopped working suddenly and stared ahead as if transfixed by some unseen menace.

She stood there for several minutes, completely unaware that Sina was trying to get her to help lift the head onto the snowy body. It was as if she'd gone into a trance, oblivious to everything around her, which scared me. Then, as if a switch had been thrown that turned her back on, she grabbed the snow head and helped lift it into place, then came straight to the house without a word to Sina. At first, I didn't know what to think about it, but then, all of a sudden, I knew. And what I knew hit me like a horrible premonition.

She stomped the snow from her boots on the porch, removed them at the door, and came straight to me. She looked like something terrible had happened, and asked if we could talk privately. I nodded, knowing exactly how she felt if what I thought had actually happened, and headed upstairs to the master bedroom. I closed the door behind us and turned to her, my mind already made up. "No, Masana. I won't go. You got a message from the Little People, didn't you? They want us to go back again, no doubt, but I won't. If you'd seen what I did, you'd feel the same."

"I can only guess at what you went through back in my time," she replied, "but this time it's different." Of course, it's different. They always say things like that to con you into helping them, but it isn't different. Oh, the time's different, the culture might be a bit different, but the negatives are always there, ready to jump out and take you by surprise. No, it isn't different. It's always the same.

"Tell me how it's different, Masana. Tell me there are no wars, no fighting and killing. Tell me there are no dangerous beasts to fight, no disasters waiting to happen. Those things are always there. I don't care what they told you, it's always dangerous. You can't believe the little bastards, honey. They'll say anything to get what they want. So, what's the story this time? What makes this one different?"

"How did you know they contacted me? They've never done that before; it's always been you they talked to. How could you have known?"

"I was watching you, honey. I saw your reaction, the surprise you felt, the almost total paralysis. I've been there, done that. I know what it's like. She—or he—whichever this one is, must not be far away. They can't transmit over long distances, you know. So, tell me what they want this time."

"Look, they took me away before it happened so I didn't see it myself, but you told me Mutasi was completely destroyed, that there was nothing left of it. But it wasn't, Mom. They said it wasn't."

"Yes it was, Masana. They made me watch, so I know for a fact that it was destroyed. And, for that matter, so was most of the world."

"But the world came back, Mom, and so did Mutasi. Parts of it, anyway. There are islands, parts of Mutasi, that came back to life."

"That's true," I admitted. "I've been there. To some of them, at least. They're beautiful. Almost as beautiful as Mutasi was."

"And my people are returning there right now. I mean, right now a long time ago. But they don't know how to live off the land and ocean. They don't know how to build fales (homes), or umus (cooking pits), or how to use coconut fibers, or weave mats, or plant crops in the jungle. They don't know any of that stuff, because they've lived in a different place for centuries. Thousands of years, maybe. That's why they want me to help. I can teach them those things. And so can you."

"Me? I don't know how to do that stuff. Remember, I was sent to a time when men did all the domestic work. I learned to be a warrior while men took care of the household."

"But you saw it done, probably even helped once in a while."

"Not enough to teach anybody else how to do it."

"That's why they need us both. I didn't do much of that kind of work either, except for the years I was with Taupulega. Think about it, Mom. Between us, we can figure it out, I know we can. Besides, it'll give me a chance to see my people again."

Well, she had me there. How could I refuse her the chance to be among her people, to languish in those fantastic islands, to speak her own language with her own people, and hopefully help them while she was at it? "Okay, honey, if they can convince me it isn't dangerous, you can go. Me, I'd be totally lost at what they want you to do."

"But they want the warrior goddesses. To inspire them, to give them trust in what we'd be doing, to let them know the gods haven't forgotten them."

"Those pricks aren't gods, Masana. They're little big-eyed bugs, not gods. And Sina and I are not goddesses, either."

"Well, that's what they called you, and what the legends of my people say, so who am I to argue? The point is, they want you. Both of you."

"They want me and Sina? Are you sure about that?" I asked, trying to keep my composure and breathe normally. Later, I would come to realize that was the turning point for me. A chance to be with Sina again was like a decadent dream, but at the moment I wasn't ready to admit it, or give in to the crazy so-called gods. "Isn't that proof that they expect trouble? If they want warriors, they know there will be a war. That's the only reason for requesting us."

"Not just you and Sina," she countered. "They also said we could bring Little Sina and Apu, and I could bring Michael. I think that pretty much suggests a peaceful mission."

Michael was her fiancé, the man she planned to marry. Maybe she was right. They wouldn't want Apu and my daughter, or Michael, for that matter, if there was the danger of war hanging over us. The request for the warriors, then, could be nothing more than what she said. Maybe it was just a symbolic gesture to inspire the arriving people, give them confidence in what Masana and Apu, who were the most capable teachers regarding life in the tropics, would be showing them.

I still didn't want to go. I mean, I wanted desperately to share time with Sina again, and it sounded like a real adventure going back with my family going along for a change, so in a way, I wanted to go. But I didn't want to see what little was left of Mutasi after the devastation I'd witnessed. It would be hard to return to islands, the tiny remnants of what had once been a proud, fantastic, and beautiful continent.

On the other hand, being with Sina again would be more than enough to compensate for that. She was more than a friend, more than just an ancient warrior, more than the woman who'd taught and tutored me in the ways of the spear. She was a kindred spirit, the sister I'd never had but always wanted.

"I'll think about it, okay?"

She nodded with that look that says, "I know what that means, Mom," looked at the floor for a minute, then turned away without another word. When she got back out to help Sina with the finishing touches on the snowman, she wasn't very enthusiastic about it.

After she left, I went to the closet and dug out my spear where it had been hidden for years. I never took it out when anyone else was around, but it hadn't been ignored. Sometimes I pulled it out just to feel the exquisite workmanship; sometimes I worked out with it, practicing what Sina had taught me. It was more than a piece of wood, that spear, carved intricately into a perfect design by self-proclaimed gods. Throughout my adventures in the past, it had been my weapon, the thing I knew I could rely on when forced to fight.

But it was more even than that. It was the tie, the bond that had been an integral part of my relationship with Sina. She had hers, and I had mine, and both had been used to fight man and beast in ancient times. There had been too many battles, too many people injured or killed, both friend and enemy. Both spears, hers and mine, had been red with blood too many times.

Nothing could make me lose that spear, not with what it meant to me. I'd keep it as long as I lived, then pass it on to my daughter, and she to her children as a reminder of those special days and nights I'd spent among The People. If the occasion ever arose that I was sent into battle again, that spear would go with me. Even if, as the stupid Little People claimed, I went back to a time of peace, the spear would go along just in case.

Yes, that spear was special to me, but what I'd been forced to do with it was anything but. I had killed and wounded men and women, all of them warriors, had helped to dispatch a saber-tooth tiger and a monstrous octopus that might have been the last of its kind. We had faced a giant at the opening to an ancient pyramid. Together, we had gone into battle too many times, and it had never failed me. It was still, after all this time, sharp and strong, capable of yet another fight if the need arose. With all my heart, though, I hoped that need would never come up again.

Was it possible this new request could actually be one in which wars wouldn't have to be fought? I hoped so, but if I went, and I knew within me that I couldn't pass up the chance to be with Sina again, I would take the spear. Maybe there wouldn't be a battle that needed to be fought and the spear wouldn't be needed, but I wasn't going back again without the weapon that had been a dependable tool and companion through two trips into ancient times, both of which turned out to be dangerous.

Just thinking about the spear and Sina made me realize that I'd already made up my mind. I would go back again. I couldn't pass up the chance to be with Sina one more time, especially if this trip actually turned out to be a mission of peace in which there were no battles to fight. The People were returning to their rightful place, the land their ancestors had inhabited for thousands of years. It was the place I had gone on my first archaeological dig, the one that turned out to be so controversial. It was there that I had been accidentally sent on my first adventure into the past, though I didn't think of it as an adventure at the beginning.

Those damned Little People were careless enough to leave behind a time travel device where they had to know someone would eventually find it. And, of course, it had to be me. Deep in a cave, I'd found two small skeletons, probably their ancestors. It was an amazing discovery, but the reckless critters stole them from me, sent them back to their rightful time, and left behind the device they'd used to do it. I found it, and it sent me on the most amazing journey of my life.

At first, I was terrified. Who wouldn't be? I'd arrived naked in what appeared to be another world and was captured immediately by a bare-breasted Amazon-style warrior when I thought such things were merely the subject of fantasy fiction. That warrior was challenged by another one of similar attire and not only lost the fight but her life in an attempt to keep me. That warrior, though frightening and devoid of common courtesies at first, turned out to be Sina. She befriended me, trained me in the art of the spear, and stood side by side with me through numerous fights. We became more than just members of the warrior caste; we became best friends, sisters of the spear.

The function of the warrior caste in the village of Tufutafoe was to "protect the forest." In other words, to protect the local villages from the Aunu'u, cannibals whose ambition was to conquer Mutasi, the continent upon which the villages were located. It also served as a forward observation post, providing info to the central government of Mutasi regarding Aunu'u activities. We, the members of the warrior caste, served and fought as a unit.

Then came the day a village woman was killed by a saber-tooth tiger, so our mission was to hunt down and eliminate the huge cat. Sina almost died in that fight, and some of the others in the caste did. It was a costly engagement, after which I had to take on an "enemy god" alone.

Perhaps my greatest accomplishment in that past was to show The People that the Little People were not, in reality, the gods they thought them to be. They were nothing more than what are often referred to as "alien greys," though they claim to be of earthly origin just like us. They have a knowledge of technology that far surpasses ours, including the ability to travel forward and backward in time, but that technological advantage doesn't make them gods or give them the ability to make decisions for people other than themselves. They're creatures just like us and as prone to mistakes as we are. But they do have the advantage of being able to see events both past and future, with which they can make decisions that would be difficult to understand for the rest of us.

The worst thing about the Little People was their knack for using people to do their dirty work. They refuse to engage in physical conflicts but don't hesitate to make humans do it for them, even when there's no benefit in it for them. Their power over humans took a serious blow when I turned on them before returning to my time. They were not invincible, and from that time forward, there was more cooperation between the humans and their so-called "gods." Apu, my lover in that ancient time, forced them to bring him to my time, where he became my husband and the father of my daughter, Sina, named after the ancient warrior and my dearest friend.

But the Little People didn't go away. Years later, they came to me again, sending me, along with Sina, on another mission to a time ten thousand years after Sina's and twelve thousand years before mine. Again, we successfully fought the Aunu'u, but in vain. Their conquest of Mutasi sparked a devastation that virtually destroyed the world. We saved the villages but lost everything. It was a horrifying mission, and I swore I'd never do anything for the Little People again. Masana was a beautiful blessing, but the only positive from that trip.

Unfortunately, they're back again. They most likely know how much I dislike and distrust them, which is probably why they contacted Masana rather than me, assuming correctly that I'd refuse. Masana, though, was a different matter altogether. She would almost surely jump at a chance to spend time with her people again, to help them if she could, and see her beloved homeland even though it was different from what it had been. And she was probably the only one who could convince me to go along. Apu wouldn't even try, and Sina, my daughter, having never had the experience before, wouldn't know enough to convince me of anything.

So they contacted Masana, and she successfully conned me into going back again. I didn't want to, but I had to. I had no wish to fight another battle, but I had a burning desire to see Sina again. So the damned little black-eyed creatures had won again. Somehow, I vowed to get even with them. I didn't know how, but I'd find a way. They were going to pay for making me go back again, although an inner voice kept whispering that nobody was forcing me. I wanted to go. I wanted to see what was left of Mutasi, and I wanted to see Sina again. No, I wasn't being forced. I could say no, but I didn't.

How far back we'd be going, I didn't know for sure, but historians claim the Polynesians first inhabited the islands about three thousand years ago, so it would probably be somewhere around there.

So there I was, ready to go into the past again. Round three coming up.

Chapter 3

I went to the backyard with Masana, where she was to inform the Little People, or person, or being, or thing, or whatever they preferred to be called, that we had agreed to go. It took me by surprise that I didn't hear the response. It was channeled directly to Masana, leaving me out of the loop despite the fact that they'd requested I go along. Or so Masana said, anyway.

Had she just said that to make me feel like I was a part of the plan but wasn't? If so, was she hoping the little buggers would accept me when she sprung the surprise on them? I was a bit hurt that they wouldn't communicate with me, but those beasts are sometimes hard to understand. Maybe they were still mad at me for turning on them twenty-two thousand years ago. That's a long time to hold a grudge, though. Then again, that's how long I'd held mine against them.

She was given instructions as to when we should gather on the back lawn for our departure into the past. I told her she should have demanded it be handled inside since there was snow on the ground and we'd have to be wearing very little in the way of clothing. We didn't want to be wearing warm clothing, at least, and definitely nothing artificial. The device couldn't transport anything artificially made. Only organic materials. No shoes, since we couldn't find any that were all organic. We'd be going barefoot, which is a bit uncomfortable in snow.

Sina and Apu were anxious to go despite my concerns for their safety. There's no way of knowing what will happen or what kinds of problems we might run into. Conflicts were always a possibility, even though we'd been promised that wouldn't happen. Then there are wild animals to consider, although I wasn't aware of any that hadn't gone extinct when Mutasi was destroyed. Sickness and disease are also frequent problems in the tropics, and they wouldn't have time to get shots against those.

Michael had refused to go, which was a severe blow to Masana, who wanted to share the experience with him. She wanted him to meet her people, see the lands she loved, let him experience time travel with her, and get a feel for the way of life she'd been raised with. I'm sure it was the time travel part that scared him off, but I can't be certain. It's enough to scare anyone who hasn't done it before.

The little creature introduced itself as Tapaosi when we were gathered and ready, shivering and freezing our butts off. That wasn't its real name, of course, since the names of the Little People are difficult if not impossible for humans to pronounce, so they used human names for our convenience. I had seen to it that everyone was fully equipped with fully organic clothing. No metal, no man-made fabrics, no zippers or buttons. Cool clothes for a hot, humid climate, which meant we were in agony while we stood waiting in the snow. Apu and Masana understood that only organic materials could be transported, but it took Little Sina a while to get it right.

"Shouldn't we pack something to eat? A snack or picnic or something?" That was Sina, who was new to the whole experience. I assured her that wouldn't be necessary, that the people we were going to meet would probably throw a feast in our honor. Then we formed a line on the back lawn, standing in the snow, and told Tapaosi we were ready. She produced the cube I was all too familiar with, and I told everyone not to look directly at it. There was a brilliant flash of light, and almost immediately we were standing on a sandy beach, backed by coconut trees and, beyond that, jungle. Nothing else.

It was a beautiful beach and warmer than the snow in my backyard, but that's all it was.

The beach was a half circle of white sand, probably three acres in size. It was a perfect place for a village to be located, but there was no village. Or any sign of human habitation. It was just a pristine, untouched beach surrounded by palms and jungle growth. Other than our own, there wasn't even a footprint in sight, and sand always reveals those.

It seemed obvious that the "gods," who were usually fairly reliable even if a bit eccentric and hard to explain, had dropped us at the wrong place. Which meant that we'd have to work our way around the island until we found the place where the returning islanders had landed. Not an impossible task, but certainly challenging, especially since we didn't know whether to go right or left. Regardless of which way we went, it was possible we could go all the way around the island, only to discover The People not far from where we started. I knew from previous experience that the island was sixty miles long and thirty-five miles wide. A long hike by any stretch of the imagination over rough lava rock and sandy beaches.

Although I loved the tropical heat with its high humidity and smell of salt water, coming to it straight from winter's frost was like a shock to our system, so we made our way from the sand to the shade of a palm tree. It didn't help a lot, but at least it got the direct sun off us, so we settled ourselves and tried to decide what to do next. There wasn't anyone around in need of the help we were supposed to be bringing except ourselves.

Before we set off in search of the people our mission was intended to benefit, we needed to find food and something to drink for ourselves. So much for the feast I'd told Sina to expect.

Coconuts could provide both food and drink if there was anybody capable of climbing the trees, something neither Sina nor I were capable of. Apu and Masana volunteered, and both made it look easy. Those of us not trained or conditioned for it knew better. It wasn't easy, and even those who were good at it sometimes fell and got injured, or worse. Husking the nuts once we had them off the tree was another issue. We had no knives or tools, so Apu went in search of a broken limb with a sharp edge we could use. A rock with a sharp corner would do to open the nut, but first, we had to get through the husk.

I hadn't seen signs of any papaya or other fruits, but if they were growing on the island, they'd likely be inland. There was a lot of vaofefe around, the plant that shriveled when touched. The People considered it a nuisance weed, but I always found it interesting. The jungle was pretty much the way I remembered it, thick and daunting under a canopy of hardwood trees. Many times I'd fought my way through that jungle with Sina and the warriors of Tufutafoe, pleasant memories that were engraved on my memory forever.

Thinking of Sina brought me to a halt. She was supposed to be with us on this mission, or adventure, or whatever it was supposed to be, so why wasn't she here? I hoped they hadn't dumped her somewhere else without bothering to tell us. I was anxious to see her again, to share whatever time this thing allocated to us. Thinking of her made me take a long look at my spear. Was I still as good with it as I'd been in my last two trips into the past? If the need arose, could I still wield it in a fight against an enemy or dangerous animal?

I was pretty certain Big Tooth, the saber-tooth tiger, was extinct, so I wouldn't have to face off with that creature again, thank God. Other than that, wild boar might be the only animal a spear would be useful for. Hopefully, there were fruit bats, wild chickens, and parrots we could catch and eat, along with crabs, lobsters, and fish from the ocean. There should also be several different fruits, berries, and tubers. Taro? I wasn't sure if taro was indigenous to the island, or if it had to be brought from abroad. The same question applied to bananas, ta'amu, and all the other basic food items.

What if there were only a few indigenous food crops on the island? Maybe pigs and chickens had to be brought from elsewhere as well. And fruit bats, white birds, parrots, and any number of other species that served as food in the other times I'd been to. What if there was nothing here but the forest, jungle, and coconuts? It'd be a pretty meager existence.

"Sina!!" That was my daughter's voice, screaming with excitement, which meant she'd spotted the warrior goddess. And sure enough, Sina emerged from the jungle followed by a grinning Apu, who carried a limb that might work for de-husking coconuts with a little bit of sharpening.

"See what I fin'," he chirped, throwing the limb to the ground and watching my daughter, who would have to be Little Sina again, rush over to give Big Sina a crushing hug. Then Masana took her turn with equal enthusiasm. There was no formal greeting between them, so that was up to me.

"I greet you, Sina, war chief of Tufutafoe and defender of the forest. It is good to meet with you again."

"And I greet you, Salini, adopted daughter of Tufutafoe, and warrior of legend. I see you have your tupa. Do you still know its use?"

"And I see you have yours, warrior goddess. Perhaps you would consent to train me again?"

"Am I truly a goddess?" she laughed, something we both found rather hilarious.

"Am I?" I asked in return, mimicking her laughter.

"It would seem that we are," she joked. "Is it not true that we have been to many places in different times, and fought battles in them all? Only goddesses, it would seem, can do such."

"Then goddesses we are," I laughed, rushing forward to hug her, dropping all pretense of propriety. It was a long hug, filled with heartfelt love and friendship that had lasted for, well, eighteen thousand years or so, give or take a week or two. When we separated, she looked around the beach and sighed.

"It is strange, this place. It is Falelima, but there is no Falelima."

She was right, I suddenly realized. It was Falelima, only there were no homes, no cookhouses, no outhouses, no people. That would all happen in the very near future, and we were there to help it happen. "The People must have gone to another place."

"Tapaosi say they come five, maybe ten days," Sina commented.

"You spoke with Tapaosi?"

Sina shook her head. "Tapaosi is stupid goddess. She don't know to talk, just think into my ears."

"Yes," I agreed with a chuckle, "very stupid indeed. Why did she bring us here before The People arrived?"

"She say we look at things, make plan for The People so we know what need to do when they come. We help them build Falelima again."

I nodded, but then a thought hit me. "Sina!" I shrieked, "if this is Falelima, then Tufutafoe is less than a day's walk from here. Why not build Tufutafoe first, then Falelima? Was it not the warriors of Tufutafoe who defeated the Aunu'u and killed Big Tooth?"

"And it was an adopted daughter of Tufutafoe, the warrior from Siata, who killed the evil god Faleolefe," she added, "and the snakes that no other warrior dared to face."

"Is it not right, then, that Tufutafoe be the first to be born again when The People return?"

She liked it but frowned. "Tapaosi say they come here."

"Then we will greet them when they come, and guide them to a better place. Tapaosi is no more goddess than we are, so it is not a serious matter to do as we wish."

"We should go to Tufutafoe and see if it still good. Maybe it is changed."

I had to agree with her logic. "In the morning, we go to Tufutafoe and see if it is the same as before."

Apu had used the sharp end of a large piece of lava rock to sharpen his limb and succeeded in husking five young coconuts Masana dropped from a nearby tree. It worried me every time somebody climbed those things. There were no limbs for support, so everything depended on the strength of the arms and legs to make the climb. Coming back down was no easier, but it went quicker and required less work. Little Sina watched Masana make the climb in awe, wondering, like I did, how it was possible.

Young coconut meat is tender and sweet, and the juice is just as sweet. Maybe it isn't Seven-Up, but to a dry throat and hot body, it's just as good. That was our dinner since we had no pots or pans to cook with and nothing to cook even if we'd had them.

Our plan had been to build a fale for shelter, but with no tools to cut the poles and framework or fronds necessary, we had to abandon that idea and wound up sleeping on the sand. It rained during the night, and Little Sina spent much of the time trying to protect her face from the heavy drops. The rest of us just took it in stride, letting the warm water fall where it may, having spent many a night under similar circumstances.

First thing the next morning, we set out for Tufutafoe. Little Sina wasn't used to walking through the jungle, especially barefoot, and found it more difficult than the rest of us, who'd plowed through thick foliage hundreds of times and knew how to navigate it without undue loss of blood. She was a trooper, though, and managed to keep up despite the pain it caused.

About an hour into our journey, we came to the pyramid Sina and I were all too familiar with. Little Sina wanted me to tell about our experiences there, likely as an excuse to let her feet rest for a while. When I mentioned the giant, or tele, which was likely extinct by now, Sina called us to a spot she was giving a lot of attention to.

There, as clear as it could be, was a huge footprint hardened in stone, exactly like a human print, only much bigger.

"That's where he stood," I shrieked. "It was muddy that day, so he left footprints that have hardened over the centuries. That means the cave should be right over here."

We found it after poking around for a while, but Sina refused to go inside. She was as impressed as Little Sina at my story about the skeletons I'd found in there, though, and we spent way too much time reminiscing. It was nearly noon when we continued on our way to Tufutafoe. Under normal circumstances, that shouldn't have been a problem, but things had changed.

The route we'd always taken before led along the base of a small mountain range, through a heavily forested area, then down a slight decline to the beach where Tufutafoe had been. That wasn't the case this time.

A little over an hour from the pyramid, the mountain simply dropped into the ocean, forming a cliff maybe two hundred feet high. It was as if the mountain had broken off and a good share of it had plunged into the depths. It was steep above the cliff, too, so we had to circle around to the backside of the mountain before we found a safe place to cross and were finally able to continue on our way.

Once again, we came across a place where the mountain fell into the ocean, but this one wasn't as steep and created less of a problem. We were starting to be concerned about whether Tufutafoe was actually going to be there. So much had changed in the horror I'd witnessed on my last trip into the past that nothing could be guaranteed.

But I'd seen both Tufutafoe and Falelima on my original trip to Samoa in my own time, before I'd gone into the past. It was there then, which meant it had to be there now. The conflagration that had changed everything had come and gone. This was the Samoa I knew from the days of my attempted archaeological dig into a pyramid that still remained unopened.

When we were within what should have been less than an hour's hike from Tufutafoe, we came across a sheer cliff that dropped again into the ocean, and there was no way across this one. We were forced to skirt all the way around the mountain, and it wasn't easy. By the time we made it to the other side, we were so winded that even Sina begged us to stop for a rest.

It was while we struggled to catch our breath that we saw it. Tufutafoe was off in the distance, just as beautiful as it had been. It faced the ocean over a gorgeous sandy beach, surrounded by coconut trees and jungle growth.

It would take work to clear a space to grow taro, ta'amu, palusami, bananas, cocoa, and other crucial crops, but it would be worth it. With lifted spirits, we set off down the other side of the mountain, and twenty minutes later we stood on the malae that would soon be Tufutafoe again. Even Little Sina was impressed by the beauty of the place. Masana didn't say anything, but you could tell by her expression that she felt like she'd come home again.

"This is where you lived and became a warrior?" Little Sina asked, trying to picture what it must have been like.

"I did, too," Masana said. "Maybe I wasn't trained as much as Mom was, but I did my share of fighting."

"And you were good at it, too," I threw in. "Together with all the people in Tufutafoe, we were able to defeat the Aunu'u. The villages were free when the destruction took them."

"And Sina, too," Masana added so Sina wouldn't feel left out. "You two were as great in battle as the legends say you were."

Little Sina shook her head. "I just can't imagine my mom being a warrior, fighting with her spear against real enemies."

“Not me,” Masana shot back at her. “It took me a long time to get used to her as a mom and teacher. All I knew of her in the past was the brave, heroic warrior who never swerved from a fight when it was called for.”

“Salini is bravest warrior I have know,” Sina broke in. “She fight Faleolefe alone, and kill Big Tooth so it not kill me. It is proud I am with her in battle.”

That brought on another question from Little Sina. “What was it like, Mom? Fighting a real live saber-tooth tiger?”

“That’s a question for another time,” I told her. “It brings back memories I’d rather not dig up right now. I’m sure Sina feels the same way.”

Sina nodded vigorously. “Very bad time. A time to not remember, but a time no can forget.” She looked around the malae as if making sure Big Tooth was nowhere around, but it was actually much more practical than that.

“We must make the fale,” she pointed out, an issue that had been on my mind as well.

“How do we do this?” I asked. “We have no knives or axes, nothing to cut the poles and thatching we’ll need.”

“It is so,” she agreed, “so first we make what we need, tools to build. Apu know how.”

“So do I,” Masana piped up. “The Aunu’u stole ours, so we had to make our own.”

Without another word, she and Apu went to work. At the edge of the malae, where the jungle took over, they started sorting through the lava rocks that were everywhere. They’d pick one up, discard it, and sort some more, find one they liked and set it aside. After a while, they’d stacked up a dozen or so, then set to work trying to use them as axes or hatchets. Most of them broke with impact, but a few held, so Apu began chipping a finer edge on those, much like early American Indians did with flint.

After a couple of hours, he'd fashioned a few crude axe heads with a little help from the rest of us, which he used to start cutting axe handles. A groove was cut into the top of the handles in which the stone rested, then tied into place with cording made from dry coconut husks, a strong cord I'd seen made numerous times before. The result was a series of axes with which small trees could be cut down and shaped for the building of a hut, or fale.

Lava is brittle, however, and some of the stones chipped away rapidly, not lasting very long. Others were stronger and did the job long enough that by nightfall the fale was beginning to take shape. We slept on the sand again that night, and first thing the next morning, after a breakfast of coconut and breadfruit, which we found growing wild not far inland, we went to work on the shelter again.

Once the poles were planted far enough in the ground and the roof framework was done, Apu started gathering thatching materials while the rest of us waded into the ocean to gather weathered coral chunks for the floor. That night we slept under the shelter of our own fale. Maybe it wasn't as perfect as those from earlier times, which were made with better tools and more time, but it was ours, and we were proud of it.

Masana went hunting fruit bats in the morning, and we had meat for a change, something we were desperately in need of. Little Sina was reluctant to eat a bat but gave in after some coaching from Apu, Masana, and nagging hunger. She was surprised at how good it was once she got past the thought. We also ate a few octopuses Sina gathered, along with three fish she'd managed to spear with a long stick. It was a good meal, but it would've been better with cocoa rather than plain water from a nearby stream or coconut juice.

A scream from Little Sina brought Sina on the run, followed closely by me, both of us with spears in hand and ready for action. Little Sina was in the fale, backing away from something that had frightened her. Sina charged in, saw what had frightened Little Sina, then did a sudden about-face and rushed back out faster than she'd rushed in. I went in, positioned myself in front of Little Sina to protect her from whatever the danger was, and then I saw it.

In the future—the future as measured from the time we were currently in, perhaps three or four thousand years from now—travel brochures would state emphatically that there are no snakes in Samoa. This wasn't that time, and as I knew all too well from my first experience in the past, that wasn't the case in this time. A boa constrictor, even one not fully grown like this one, is a big snake. It frightens even people like me, who are familiar with American snakes that tend to be much smaller. Both Sinas, however, were more than just frightened; they were terrified.

The second I stepped in front of her, Little Sina tore out of the fale to join Big Sina at the ocean's shore, which was as far away as they could get. "It is a demon," I heard Big Sina shriek hysterically, a reminder of the first time I'd faced such a creature. This one was smaller than that one, thank goodness, but it was still imposing. This time, though, I had my spear, which gave me more confidence. In our little retinue, there wasn't anyone else who would come anywhere near that thing, so again it was up to me. Shades of a day long gone.

"You shouldn't have come here," I told the creature in a soft, almost breathless tone. "If you'd just left us alone, I wouldn't have any beef with you, but now you've invaded our home, and I won't have that. And don't try that old 'but I was here first' routine, either, because I don't give a damn. We built this fale, so it's ours, and you weren't invited. So, either you run for your life, or I'm going to have to kill you. Which will it be?"

Well, it didn't run. Not that snakes run, anyway. The damn thing just lay there on the floor looking up at me as if to say, "I'm staying, so what are you gonna do about it?"

"You don't think I'll do it, do you?" I asked. "Look, I killed your great-great-great-grandfather a thousand times removed, and I'll do it to you, too, if you force me to." I don't know what Sina, Little Sina, Apu, and Masana thought I was doing, talking to a snake like that, but I knew. I was buying time, trying to build up the courage to start the inevitable brawl.

My mouth was dry, my lips parched, sweat was beading on my forehead and on the palms of my hands. I was scared, and smart enough to realize it. This snake was on the floor of our fale, not perched in a tree with limbs separating us. With a sweep of its tail, it could take my feet out from under me and wrap me in its deadly coils. It was big enough to do that, I figured. I could see no bulge indicating it had eaten recently, which meant it would likely be quicker than the one I'd fought before. The only thing that was the same with this fight and the last one was that no one was going to help me.

I was on my own.

Again.

"I wish you'd just skedaddle," I went on. "Shoo. Go on, you ugly thing. I won't be able to catch you in the jungle, so if you stay there, I won't bother you. As long as you don't bother us, that is. Can we make a deal?"

Nothing. That stupid creature wasn't even listening, but it wasn't advancing, either. It just kept looking up at me with a blank look on its face, its tongue slithering in and out. Maybe that was to sense whether I'd make a suitable meal or not. I didn't know what was on its mind, but I was left with no choice but to make the decision for it. I lifted my spear, pointing the tip directly at its face, and took a step forward.

"See that? See that, you moron? That's what I'm going to kill you with if you don't get the hell out of here." But it didn't move, or change expressions, or anything else, so I took another step forward. I was within maybe four feet, but it still didn't move. Another tiny step brought me to within three feet, and still no reaction. At that range, I could reach it with my spear, but I didn't. It was still far enough away that a solid, killing blow would be difficult.

I needed to drive the point of my spear straight down into the head of that beast, which meant I'd have to get right up to it. But that also meant it could grab me all too easily. I hesitated. And when I got done hesitating, I hesitated some more. These creatures could be fast. I knew that from prior experience and from a dozen videos I'd seen. Then there was the question of which way would be more effective at killing it. The point of the spear, driven into its skull, or the hilt to beat it to death?

Finally, it made the decision for me. It started slithering toward me, still without any discernible expression on its face. It was a slow move but intentional, and I reacted instantly. I let out the war cry I hadn't used in a long time and drove the point of my spear at the top of that evil head. It was a hard, well-aimed drive, solid enough to kill anything.

It bounced off the skull of that ugly monster without penetrating anything deeper than the skin. It was most certainly not a killing blow.

The snake recoiled, the head pulling back but not the body, which quickly coiled or wrapped around itself. It was now only two feet away, and its body was laid out like a loose coil of garden hose. For the first time, I could see the full length of it, maybe eight feet or so. About half as big as the last one I'd killed, but big enough. The mouth opened as if to strike (rattlesnakes do that, and they were the snakes I was most familiar with), so I jammed the point of my spear straight into that open orifice.

This time it penetrated. The snake's head recoiled farther away, but the body started coming at me, like a rolled-up rope crossing a rough floor. I took a quick step back, pulling the spear free from the now bloody interior of its mouth, and with a wide swing, brought the blade edge down behind the head. It cut deep, not quite slicing the head off, but more than enough to kill it. But snakes aren't endowed with much intelligence.

The damn thing didn't know it was dead and kept rolling those coils closer and closer to me.

I jumped out of its reach and again brought the blade down in a vicious swing. It hit a different spot and again failed to chop off the head, but it didn't matter. The thing was dead, so all I had to do was leave and wait for it to come to that realization. I backed out of the fale slowly, watching it carefully as I went, wary of any tricks it might play, then fell apart.

It was always that way with me. I'd know fear when confronted with danger, but keep my composure and face it with whatever was needed. When it was over, though, I'd lose it, start trembling and shedding tears. So there I was, the fearless warrior goddess, shaking violently though the danger had passed, unable to stop. For long moments, the snake squirmed and rolled in its death throes, which eventually got slower and slower, and finally stopped. The killing part was over.

"You did it, Mom!" a voice shrieked behind me. It was Masana, the first to come anywhere near, although she was still a good twenty feet away. "You killed it."

"I didn't want to," I cried, my chin quivering, "but it wouldn't go away like I told it to."

"What good would that have done? It'd still be out there, and eventually we would've had to deal with it again."

"Yeah," I sighed, trying to gain control of myself. "Maybe you're right. Anyway, this part's done." Then I turned to Sina, who was still standing just above the surf. "Does it have to go to the ocean, Sina?"

"It must be eaten by creatures without legs," she shouted back, "or it will come back and do evil."

"Damn!"

"I've never heard you swear so much," Masana said with a nervous giggle.

That stopped me cold. "Was I swearing? I don't swear, and you know it."

"But you did, Mom. A lot."

"No, I didn't. If I did, it was that damn snake's fault. I don't swear, and you know it."

"I think it's cool."

"Swearing isn't cool, Masana. It doesn't matter why. I'm ashamed of myself."

"I meant it was cool that you dared to kill it."

"Oh, that. Well, it had to be done, didn't it?"

"And you did. Wow, just look at the size of that thing."

"Yeah. Trust me, I did."

"Salini is brave warrior. She kill the demon, like before." Sina had come to within thirty paces but clearly wasn't coming any closer.

"And thanks for your help both times," I chided her. "Will you help me get it to the ocean?"

Her answer was to back off another ten paces.

"How about you, Masana?"

"I'm not gonna touch that thing," she groaned.

“I’ll help, Mom,” Little Sina piped up, “but why does it have to go to the ocean?”

“Because of what Sina said. They believe if it isn’t eaten by creatures without legs—in other words, fish—it will come back to terrorize the village. Or some such thing. Anyway, it’s their belief, so we have to honor it. We’ll have to drag this thing out to the water, then sort of float it out to deeper water.”

“Sounds crazy to me.”

“Don’t let Sina hear you say that. Or Apu. They have an unbelievable fear of snakes and regard them as inherently evil.”

“I gathered that. How heavy is it?”

“I guess we’ll find out.”

It was heavy, but once we got it on the sand, it wasn’t too hard to drag. The ocean was a bit choppy, but we took the critter out until we were almost neck deep, fighting the waves all the way, then finally let it go. It sank, and the undercurrent seemed to be taking it farther out, so our job was done.

Back on shore, though, we found Sina making Masana pick out all the coral pieces on the floor of the fale with blood on them and throwing them into the jungle. Then they worked on smoothing out the holes they’d created. The result was a thinner floor, but it still worked.

Little Sina wondered about that. “Why do we sleep on coral, Mom? It doesn’t make sense.”

“Actually, it does, but not just any coral. Fresh coral has sharp, jagged edges, but old dead coral has been smoothed by the water, like river stones back home. And it drains immediately, so even if it gets wet, you don’t have to lie on a soggy surface.”

“So why not rocks, then? Wouldn’t they do just as well?”

"Small river rocks would, yes, but I haven't seen any of those around. Have you?"

"No, I guess not."

"Most of the rock here is lava, which is also covered with sharp, jagged edges. It would drain fine, but you'd never get any sleep. Another advantage of coral is that it shifts with your body's weight. With the mats Apu's making, it'll be like sleeping on a mattress."

"I doubt that."

"You'll see," I promised, putting an arm around her shoulders and guiding her toward Sina, who was beckoning us with an arm.

"We must return to Falelima, for The People may arrive soon. But we have much work to do here if we are to show them how to live."

"Then we will split up. It is Masana, I think, that they expect to greet them, so she should go to Falelima. And I'm not much help at making a fale or making preparations for the building of a village, so I will go with her. You and Apu should stay here and continue your work."

"Little Sina should go with you as well."

"If it's okay, Mom, I'd rather stay and help Dad. I didn't care much for the hike over here. My feet are cut and sore, and besides, I'm really interested in how things are made in this place."

"Are you sure?"

"I'm sure I don't want to cross those cliffs again."

"Then Masana and I will go, and when The People arrive, we will bring them here by boat, for the inland trail would probably kill some of them."

"It is agreed," Sina nodded.

"It is agreed," I nodded in return. "We will leave in the morning."

There was something else bothering Sina. "What do we do if another snake comes?"

"Kill it," I told her. "You saw how easy it was."

"It is easy for you," she replied, a look of abject fear in her eyes. "It is something I cannot do."

"Of course you can. Your spear is as true as mine."

"Then it is something I will not do," she amended.

I nodded my understanding. "It is my belief that there are no others in this area, for I think they are not friendly to their own kind. If one were to invade the territory of this one, it would probably eat it, for they do not share territories, like Big Tooth did not share with others of its kind." The abject fear in her eyes turned to terror. "No, Sina. There is no Big Tooth in this time."

"Is this a truth, or do you guess?"

"It is a truth. They became no more many thousands of years ago, when Mutasi also became no more. Of this, I am certain."

"But Mutasi remains, so maybe Big Tooth remains also."

"Mutasi does not remain, Sina. Not as you knew it. You saw the changes coming here from Falelima. Parts of it remain, but not all, and Big Tooth died many thousand cycles ago. It is so."

"And yet the serpent remains. How is this so?"

"That I do not know, but I think there will not be many."

"As long as that be so, I will remain with Apu and Little Sina while you and Masana go to bring The People."

"It is good. We will go in the morning."

Chapter 4

There was a heavy rain throughout most of the night, so Masana and I slept under a coconut tree that provided little shelter. We were drenched, but in the morning the sun came out and the skies cleared. It was hot and muggy, but the sand dried out about as fast as we did, maybe even faster. We munched on some tubers Masana found and cooked the night before, since neither of us were in the mood to try starting a fire, or cooking if we succeeded. The food was wet and cold, but edible.

We'd started building a fale but didn't have the foundation done yet. Once the poles were set, we'd have to build the rafters that connected to support a roof, then cover it with thatching. That was our goal for the day once the water ran off the trees, but as plans usually go, ours got interrupted.

Masana spotted it first. "There, on the horizon," she shrieked. "Isn't that a boat?"

I peered in the direction she pointed, shielding my eyes from the sun. "Yes, I think you're right. It looks like a sail to me." An hour or so later, we could see several, all coming toward us and growing bigger as they drew closer. "It looks like The People have arrived."

"My people," she said, excitement emanating from her words.

"Yes, your people. They've come home again at last."

"What will they be like, Mom?"

"We'll find out soon enough, I guess, but I doubt that they've changed much."

At a distance, they looked small, just a dark line on the horizon. In the ocean, even an island can be missed from a mere forty miles away, which meant these boats, much smaller and lower than an island, must have been no more than ten miles out, maybe less. As the distance closed, we could make out sails first, above a black line. Not the white billowing sails of the tall ships that would discover these islands a few thousand years in the future, but squat, squarish sails that didn't reach far into the sky. And there were dozens of boats, all of the same design.

"This is exciting, Mom. I can't wait to meet them."

She was excited, no doubt of that, but there was also a note of apprehension in her tone. I shared that with her. Would these people turn out to be like the Polynesians we knew, or a different sort altogether? Did they still speak their native language, or would communication be an issue? After all, they'd lived in a different part of the world, probably somewhere in Asia or Indonesia, for untold thousands of years. How much of their original culture remained?

I'd expected the boats to be large outrigger canoes, bigger than the ones I was familiar with, but not huge. I was proved wrong as they came into full view. They were fashioned along the lines of outriggers, alright, but so much bigger that it took me by surprise. There were three extremely large canoe-like bases, like a tri-hull boat, but over them was what could almost be called a small village. A deck had been erected over the canoes, with two sails, one in the front and another in the back, suspended on wooden masts. Between them, a thatched roof provided shade and shelter for what must have been several dozen people.

It wasn't just people, either. At the very back, behind the rear sail, was a series of cages in which pigs and chickens and a few other animals I couldn't make out yet were kept. These, I assumed, would be the progenitors of the wild boar and wild chickens that would populate the islands in later times.

How these people had navigated the huge Pacific Ocean without getting lost was a mystery. Did they use the stars to guide them, or the currents, or did the Little People show them the way? And how could they have stored enough food and water to last them for the months they'd been at sea? Did they make stops at other inhabitable islands on the way? If so, why hadn't they stayed there rather than coming all the way to what would become Samoa?

We waited for over two hours before one of the boats crossed the reef, which was possible because it was high tide, and came to the shallows just offshore. There it stopped. A single man was seated on a dais-like bench at the very front of the boat. He was huge, probably well over three hundred pounds, and ornately covered in clothing of bright colors. Some of the colors were actually bird feathers attached to his outfit, others artificially dyed fabric. Two young men, naked except for a breech clout, rushed forward to help him stand as if he couldn't do it himself. Given his size, I wasn't sure he could.

Once he was upright, though, he proved to be quite agile and capable of handling himself. There was a trumpet blast from a conch shell as he stepped from the boat into the water that came up nearly to his hips. He plowed through the surf easily and finally dragged himself onto the sandy beach, where we awaited him.

"Talofa lava lau afioga," I called out in a firm voice, hoping he understood the traditional greeting and was indeed the chief it was meant for. For a long moment, he just stood there, looking me over as if puzzled, then grinned.

"Talofa lava fa'afine." The initial part of the greeting was now done, but the formal part was yet to come.

"I would greet you properly, but I know not your name or title. Is it allowed for me to know these things?"

"It is allowed," he nodded. "I am Petaio, high chief of Sa'atofu, first village of The People. My talking chief will come to present a formal greeting to you if I may know your name and title."

“I am Salini, adopted daughter of Tufutafoe, warrior and defender of the forest. Beside me is Masana, princess of Mutasi and daughter of Queen Sinasi in days long gone.”

Petaio dropped to his knees as if struck by a battle axe, then gave me a pleading look. “I apologize for standing before a goddess and princess. It is not right, and I will do it never again. Please forgive my ignorance.”

“Rise, Petaio. I am a warrior, not a goddess, and Masana is the princess of a land that no longer exists. We are like you, a part of The People. Nothing more.”

“But the legends tell of the white warrior from Tufutafoe, that she is a goddess of war along with her friend Sina.”

“The legends are kind, but stretch the truth. Me and Sina are people, like you and your companions on this journey. It is good that you have arrived well and safe.”

Petaio got to his feet easily but tentatively, as if he wasn’t sure he should. His arms and legs, though obscenely obese, showed strong muscles and sinews. He was big, but he wasn’t all fat. “I thank you for your greeting, Salini of Tufutafoe. And you, Masana, princess of Mutasi. You are also spoken of in the legends, taken by the gods at the time of the great flood that destroyed our homeland. Now we are back to rebuild on this, the remains of our traditional homeland. If it please you, Princess, we will call this island Savai’i, for it is our traditional home.”

I motioned for Masana to respond, for his question had been for her, and although she felt a bit overwhelmed, she managed a nervous response. “It is a good name you have chosen, Petaio. It shall be as you say.”

“I am glad, Princess. Now, then, I will bring my talking chief and retinue ashore to give you a formal greeting. And then we will begin to build homes and cookhouses, and make our village.”

"It would be better to wait just a short time longer, Petaio," I broke in. "A short distance down the coast is a better place to build. It is where the Tufutafoe of legend was located. To build there is to become a part of your history, to make your homes in the place where The People once did great deeds. You and those with you will make Tufutafoe great once again."

"Is this place not good?"

"It is good. It is the place where Falelima stood in times now gone. But Tufutafoe, I think, is better."

"We have traveled for many moons, Salini. We have a few who are not well, women with child who are weary of traveling, and we are anxious to put our feet on the ground once again."

I nodded my understanding. "I hear your words, Petaio, but Tufutafoe is but a few more hours by boat, and worth the time. Sina awaits us there with Apu and my daughter. Can you endure just that much longer?"

"Sina is there? The warrior goddess?"

"Yes, Sina, the war chief of Tufutafoe, defender of the forest. She waits for you, Petaio. Will you go to her? You and your people?"

"It is a good thing. Yes, we will go."

At that moment, another boat pulled in alongside Petaio's, and a very old, wiry thin man jumped into the shallows and waded ashore. Despite his age, he was full of energy and vigorously strode to us.

"We are here, Petaio. Our journey is over."

"Talofa lava lau afioga," I once again said, using the traditional greeting for what I perceived to be a high chief. For a moment, the newcomer paused, then turned to Petaio. "Who is this woman who speaks to me?"

“She is Salini, the warrior from Siata, adopted daughter of Tufutafoe and defender of the forest. The one the legends speak of.”

“Is it true?”

“It must be true, for she told me so herself.”

“Then it must be true.”

“I am Salini of Siata, adopted daughter of Tufutafoe,” I said to break in. “I would greet you properly, but I know not your name or title. Am I allowed to know these things?”

“If you are truly Salini of Siata, you are allowed.”

“I am truly Salini of Siata,” I replied.

“See, she carries the spear of legend, of which there can be no other,” Petaio broke in, sticking his face to within inches of the newcomer’s.

The newcomer stood to his full height and turned to me. “I am Masotua, high chief of Amasiala.”

“Welcome, Masotua, high chief of Amasiala.”

“My people will come and make a feast. We will have kava to celebrate the end of our long journey.”

“This should wait but a little longer, Masotua. A better place is but a few hours sailing from here.”

Petaio broke in at that point. “No, Masotua wishes to stay here. He and his people will dwell in this good place, and me and my people will go to the other place that is not as good.”

The two chiefs started bickering back and forth, obviously not the best of friends, while yet another boat approached the shore and still another man dropped to the water and waded ashore. This one was young and strong of build. For a moment he studied the two chiefs arguing, then turned to Masana.

“What do they fight about, girl?”

“Whether to stay here or go on to a different place,” she replied, somewhat irritated at the way he’d addressed her.

“What is wrong with this place? I see nothing to cause concern.”

“There is nothing wrong with it, but there is a better place not far away. I would welcome you formally, but I know not your name or title. Am I allowed to know these things?”

“No,” he said. “I do not tell such things to a common girl.”

That brought me into it. “She is no common girl. She is the daughter of Queen Sinasi of olden times, princess of the Mutasi that was, and you will address her with respect.”

“Is it you who will force me to do such?” he said, turning to face me for the first time, but then his eyes went wide and he choked on whatever else he intended to say. He focused on my spear as he spoke, his voice soft and respectful for a change.

“That spear says to me that you are Salini, the warrior goddess. Is it so?”

“I am Salini, the warrior from Siata and adopted daughter of Tufutafoe, defender of the forest.” I was getting tired of saying that. “And this is Masana, princess of Mutasi, daughter of Queen Sinasi. She would welcome you formally, but she knows not your name and title. Is she allowed to know these things?”

“Oi! Yes, my princess,” he said, quickly turning back to her. “I am Atamiso, high chief of Kusani.”

“I welcome you, Atamiso, high chief of Kusani,” Masana mumbled in barely controlled anger.

I stepped forward. “There are three high chiefs who have taken this journey?”

“Yes,” he nodded, not overly burdened by intelligence. “And more. I think there are seven. Or maybe more than that.”

“And each is the high chief of a village?”

“We are the chiefs of our families.”

“You are family chiefs? And there are at least seven of you?”

“Or maybe more,” he added.

“You don’t know how many there are?”

“I don’t know to number them, Salini of Siata. Is it important to know this thing?”

“Who among you is in charge? Who is the highest of the high chiefs?”

“We are the chiefs of our families.”

“Then which family is biggest?”

That got me a blank stare. I needed to find out who made the decisions for the expedition, and Atamiso didn’t seem capable of giving me that crucial bit of information. I went back to where Petaio and Masotua were arguing and broke into their debate.

“Esteemed chiefs, forgive my interruption, but would you tell me who is the leader of your journey?”

“I am leader of the family Sa’atofu,” Petaio said rather defiantly.

“And I am leader of the family Amasiala,” Masotua boasted.

“But who is leader over all the families?”

“No other chief leads Sa’atofu,” Petaio demanded. “I would not let it be so.”

“No other chief leads Amasiala, for I also would not let it be so,” Masotua added, not to be left out.

“But there must have been one chief who made the decision to come. Which chief is that?”

While I was asking that, another boat with another chief arrived.

“We all decide to come, and lead our families to this place,” Petaio explained. “All families say yes, we go to traditional homeland, so we go, and now we have come.”

Frustrated, I turned to give the traditional greeting to the new arrival as still another one began wading toward the beach. That’s when I gave up. “All chiefs must listen to me, Salini of Siata, adopted daughter of Tufutafoe. You must return to your boats, and we will guide you to the location of Tufutafoe, a respected village of legend, where you will build your homes.”

Petaio was quick to argue that one. “Sa’atofu will have own village, not share with Amasiala or others. We go to Tufutafoe, they stay here.”

That opened up a real can of worms that I thought might break out into open warfare. None of the families wanted to share their village with any of the other families, but they all thought they should be the ones to rebuild Tufutafoe. There was a lot of screaming and shouting and pointing fingers and angry dancing about the malae, making me grateful none of them brought weapons ashore with them. They each wore, however, a whip-like device that was worn over a shoulder, part of the traditional chief’s garb, and they switched each other time and time again, the anger growing more and more intense by the minute. The switches might not have been weapons, but they sure did function like one.

I had to scream for silence to get their attention. “Listen to me, all you chiefs. Your families may differ, but you are all of The People. You are the rightful inhabitants of this, your homeland. After this great time, you have returned to where you belong, so it would be good to set aside family disputes and unite as The People.”

Nope, that didn’t strike a chord with any of them, so I proposed a council of chiefs to discuss the matter further. Akimatolu, one of the later chiefs to have landed, added that there should be a feast.

“And dancing,” Petaio said, expanding on the theme. “My family is much good dancing.”

“And mine,” Akimatolu exclaimed. Two other chiefs chimed in, claiming their families were the best at dancing, so I finally suggested that each family perform a couple of dances.

“We will bring kava,” Atamiso proclaimed, which took me by surprise, but pleasantly so. Kava was a ceremonial drink that I had come to like in my trips into the past, and it had a calming effect on those who drank it. It is a muscle relaxer that makes one mellow after a few shots, but not drunk. Maybe it would help the chiefs come to an amiable solution.

By early afternoon, the malae was filled to capacity with the families that were glad to be on solid ground again after months at sea. The boats virtually filled the coastline in front of the village, some behind others, so the people on those in the back nearly had to swim to shore. Hastily built cooking pits were made, and fresh pork was put on to cook along with taro they’d brought with them and some roots dug from the jungle floor. Some fish were caught, a few octopuses were captured, and over two dozen fruit bats finished off the preparations.

Polynesian dances are traditionally done at night, when it’s cool enough to enjoy, and performed in the light of torches. Although these were disparate families, they all seemed to understand the protocols of the kava ceremony, which got underway a couple of hours before sunset. They wanted to serve me first, which is the honor given to the highest-ranking among them, but I declined in favor of Masana, whom I introduced as the true princess of Mutasi, the traditional homeland of The People.

The kava ceremony went smoothly enough, although there was a brief dispute over who would be the taupou, or village virgin, though they rarely are. The taupou serves the kava topless, except maybe for a flower lei, but there weren’t any flowers around to make one, so Tipiana just served topless. The chiefs were amiable enough about who got served after Masana and me, but then the argument over who would occupy Tufutafoe got started again.

It paused when the feast was served and continued unabated until after the dancing was done. These were a little less stimulating than the dances I'd seen in more ancient times, but still stirred the heart. Then the argument fired up again and went well into the night. By morning, it was decided that the family Sa'atofu, led by Petaio, being the first to have made land, would go to Tufutafoe. Masotua and the family Amasiala would stay at what would be Falelima, and the rest would sail up the coast in search of suitable places in which to build their villages.

That was going to complicate our mission. We were supposed to teach them how to live in this climate and environment, which would be difficult with them scattered all over tarnation, but it seemed we didn't have a choice. Those damn Little People had given us an almost impossible task once again, and there wasn't much we could do about it.

Trouble started again in the morning when everybody was getting loaded up and ready to leave. Petaio insisted that Masana accompany him to Tufutafoe, for she was no longer a princess because her mother had died. She was now the queen of Mutasi. Or what was left of it. The other chiefs objected, saying that since he was getting the best location, Masana should help those in less desirable places.

At first, it was just an argument, but then it escalated into a shouting match. Even Sinaloa, the head of the Moana family, the only female chief among them, got into the screaming with her high-pitched, screeching voice. She was, surprisingly enough, the first to declare war on Petaio, threatening to take up arms—her and her family, if necessary—to overcome Petaio's claims to the queen.

Masana was terrified, shrinking back from the fray. She didn't want to be the cause of trouble but had no idea how to prevent it. It was when three stout men of the Moana clan ran forward with makeshift clubs that all hell broke loose. The Sa'atofu clan scrambled off to get their own weapons, this time knives and spears retrieved from their boat, so the Moana men rushed to their own vessel to get theirs. War had been declared.

This wasn't the beginning I'd expected, and it certainly didn't fit into the promise that ours would be a peaceful mission without conflict. But there it was, so I had to do something to prevent it, and I had to do it fast. What I did was brash, probably, and obviously insane, but it was the only thing I could think of. I only hoped it would work.

Rushing to the center of the malae, I let out the fiercest war cry I could manage, raised my spear high, and shouted loud enough for everyone to hear. "There will be no fighting," I screamed, "for I, Salini, adopted daughter of Tufutafoe and warrior from Siata, have so said. Anyone who would fight will fight me first. This is my word. Put down your weapons or raise them against me if you dare, for it is I who killed Faleolefe and Big Tooth. It is I who killed the Aunu'u who had captured our villages, and it is I who will kill any among you who wishes a fight."

For a moment there was silence, and then Petaio laughed. "You are brave, warrior, but we are many and you are one. I have twenty warriors, and they have known battle before."

At that instant, Masana rushed to my side, holding her own crude spear and a knife she had wrenched from the hands of a man near her. She stood next to me and cried out in her loudest voice, "Now there are two," she began, "and I also killed the Aunu'u along with Sina and Salini. I too have been in battle. Send your twenty, Petaio, and you will have twenty less when we are done with them."

I'm not sure what went through Petaio's mind in that moment. Maybe he believed we could really defeat his warriors, or maybe he just didn't want to fight against the woman he considered to be the queen. Fighting me wouldn't have been out of line, for I am a warrior, but fighting the queen was something else entirely. Even if he managed to kill us, the other families would kill him for his treachery. But it didn't matter what went through his mind. After a short pause, he ordered his warriors to drop their weapons. All of them did, except one.

He was a young man, maybe twenty years old, tall and strong. His hair was long, like all of them, but combed, unlike most. There was menace in his eyes, a wild look such as I'd seen too many times. The spear he carried wasn't fine and polished, but it was sharp and battle-ready.

"I will not drop my weapon at the word of a woman," he screamed, **"even one who thinks herself a warrior. I will fight you, Salini, and I will kill you, but the queen must not fight."**

"You think to fight me alone?" I hissed at him. "Your companions have seen no wisdom in continuing to fight, and will not support you. You would be one young man against a warrior who has had more blood on her blade than you have ever seen. Do you wish a quick death, warrior, or should I play with you for a time? This I promise you; if you cross the malae with your spear in hand, I will certainly defeat you."

"No woman threatens me and lives," he screeched, his challenge blocking out all other sounds. Petaio was shouting for him to drop his weapon, but he either didn't hear him or ignored the command.

"And no man has raised a spear against me and lived to walk away, boy. Put down your spear and let there be peace between us. I have no wish to kill you."

"It is I who have a wish to kill, supposed goddess, and it is you I will kill." Without another word, he sprang forward, running hard across thirty feet of sandy beach.

I shoved Masana away in order to prevent injury to her and give me the room I'd need for such a fight, then raised my spear to await him. I held it over my left shoulder, the wrong direction for a strike in the direction he was coming from, but it was clear to me that he was awkward and unskilled. His spear was over his right shoulder, and he clearly intended to swing at my head, which gave me the advantage. His swing started early and was well into its arc before he reached me. Seeing how it was coming, I ducked under it and swung my own spear as he passed, arcing toward his back as he went by.

The teeth of my spear dug in just below his shoulder blades, deep enough to make him fall to the sand in pain, but not enough to cause permanent damage. His spear went flying from his hand and landed ten feet away. He was lying face down in the sand, writhing in pain as I came to stand over his weapon.

"Your boy thinks himself a warrior," I called out to Petaio, **"and he is brave enough to become one when he learns the proper use of his spear, but that is something he does not know yet. Now you must tell me if you wish him alive or dead, chief of the Sa'atofu? I will honor your demand."**

The boy squirmed in the sand and, despite the pain that was evident on his face, turned to his side and looked up at me. **"I would live,"** he wheezed.

"It was not you I asked," I shot back at him. **"Petaio will decide, for it was he you disobeyed. You are brave and strong, but foolish. Did I not tell you it would be foolish to attack? Did Petaio not tell you to put down your spear? Did you heed my warning, or do as Petaio told you? No, it is your chief now who will decide this thing that must be done."**

"I will... obey him, for that is... my word," he said, struggling for the strength to speak.

"If he chooses to let you live, I will see that you keep your word, boy. What is your name?"

"I am Masito," he gasped as another wave of pain swept over his back and shoulders.

"Then this is my word, Masito. Your spear is a poor one, not suitable for hunting or fighting, and you know not its use. If you live, come to me when you are well again, and I will show you how to make a good spear, and teach you its use."

"I will do as you say," he groaned.

"If you live," I added.

"He will not live," Petaio called out. **"He has attacked and offended the great Warrior from Siata, and shamed my family. For this, he must die."**

Masito's face fell, his gaze on the tip of my spear, still held over him. The tip was sharp enough to pierce his heart or split his lungs, the blade sharp and strong enough to sever his head, the teeth capable of crushing his skull. My spear had done all of those things in times past. I held it over my head, pointing the tip at his chest and kept it there for what must have been an eternity to him, then lowered it to my side.

"If Masito is to die," I told Petaio, **"it will not be at my hand. He is a brave man and will obey you from this time forward, for that is his word. I kill in battle when needed, but I do not kill to please a chief. If you have a healer, let her see to this man and make him whole again."**

Chapter 5

After the skirmish with Masito, no one challenged my word again, so I proceeded to make the decisions from that point forward. Even as Masito was being carried to their boat for treatment, I was giving orders on how things would be done.

"Masana will go with you, Petaio, and will oversee your work at Tufutafoe. Sina, the warrior chief of Tufutafoe, will assist and protect her, for I do not have much trust in you, chief of the Sa'atofu. For ten days, they will stay with you, for this is long enough to build a village. You will also build an outrigger canoe big enough for five people. In it, you will put five paddles, and in ten days Sina and Masana and Apu and my daughter will come here in that canoe.

"For ten more days, we will assist the people here at Falelima to build their village, and then we will go to where the next family has chosen to build, and assist them, and then on to the next. Ten days for each family, and then our work will be done, and you will fend for yourselves from that day forward. That is my word. Is there anyone who will not abide by it?"

There was no opposition expressed, so Masana and the family Sa'atofu finished loading their boat and set sail for where Sina, Apu, and Little Sina waited. Masotua gave instructions for the unloading of his family and their goods while the remaining families reloaded theirs in preparation for departure.

"There are some good places to settle in this direction if I remember correctly," I said, pointing to the south, **"but those the other way are far beyond Tufutafoe and are mostly rocky cliffs. It is best to go as I directed, but it is for you to decide. Wherever you go, we will find you and come to assist you. Do you accept my word?"**

They did, and by midday they were on their way. In those boats, every one of them should have been able to find a suitable site before dark if my memory served. There had been dozens of villages in that direction when I first came to Samoa, but north of Tufutafoe the terrain was rugged. Cliffs dominated the coast, with steep mountains and thick jungle behind the beaches, which were small and rocky. What these people needed was a good, sandy beach to build around, with low sloping hills beyond that could be cleared for planting taro, ta'amu, bananas, oranges, and a myriad of other tropical foods.

It was going to take some time in the outrigger Petaio was supposed to build for us, but we'd be able to find the newly developing villages without trouble. The boats were so big there was no way we could miss them where they were moored, and almost all villages are built on the coast, usually surrounding a sandy beach, which they called a malae. It was sort of the civic center of the village, belonging to everyone. There was no word in their language for personal ownership. Everything belonged to, and served the needs of, the village.

Even children were, to some degree, village products. A nursing mother might be called upon to feed another woman's child, or give milk for medicinal purposes to anyone in need of it. If a woman was unable to bear children of her own, a woman with several might give her one or two. Likewise, those who tended the crops did so for the village, not just their immediate family. Coconuts would be harvested for all, and hunters brought back wild boar, fruit bats, or chickens for the village to consume.

It was an outstanding life if you could make it work, but there are always greedy people who ruin things for everybody else. More often than not, it was the chiefs who screwed things up, because they thought they were entitled to more than the untitled by virtue of their status, like politicians in my time. Usually, it was the high chiefs, but sometimes lower-ranking chiefs like the talking chiefs. In some families, there was more than one high chief, ranked in descending order from the highest on down. There was generally only one talking chief in a village.

It was the talking chiefs who handled ceremonial events and narrated the legends and histories that were handed down orally from generation to generation. It was the duty of the talking chief to memorize those legends and histories, and even when an event didn't call for such orations, the respect language was largely built on them. To fully understand the respect language, you had to have at least a fundamental knowledge of the legends, myths, and histories of The People.

Some of the legends and myths passed down were familial in nature, so in that case, the talking chief would be talking about the ancestors of his own family. Most, however, encompassed all the families of The People. Those were the ones that Sina and I were a part of, regardless of family connections.

The histories were the most intense part of the talking chief's obligations. It went all the way back to the creation of humanity, which they claimed took place in Mutasi. There was no room for evolution in their history, insisting that it was clearly a creation by beings who came in amazing flying machines and took it upon themselves to create us. Sumerian texts tell a story that is surprisingly similar.

The histories also told of the splitting off of factions from the original creations, such as the Aunu'u and dozens of other peoples. There were differently colored people, or races, all created in the beginning, and many of them left Mutasi to build their homelands in other parts of the world. There were those who were black, yellow, and white, like me, but they were all part of the original creation in Mutasi.

The myths were a little more theatrical. These were stories that tended to be more mystical, and therefore less believable, but they were accepted as truth by the orators. One such myth was that of the turtle and shark. It's the story of an old woman who was frail enough that it was difficult for her to walk. After a hurricane destroyed the crops, the chief had his people gather all the food they could find before it spoiled, and throw a final feast, which was held at the malae.

Because the old woman couldn't get there, she sent her granddaughter, who was caring for her, to bring back some food. The chief refused to give her any, so in anger, the old woman had the granddaughter help her to the edge of a cliff overlooking the ocean and cried out in anger to the villagers. Then she jumped off the cliff, and on hitting the water turned into a white shark. The granddaughter jumped after her and turned into a multi-colored sea turtle.

Although it seems beyond the realms of logic, that village still, when I first came to Samoa in my own time, went to that cliff daily and chanted an apology to the old woman and her granddaughter. And when they did, an unusually white shark and a brilliantly colored sea turtle swam to the shore as if to accept the apology.

There were other stories just as incredible, dealing with shape-shifting and unnatural things. Were they just fables meant to emphasize the importance of humanitarian acts, or based on a truth that defies logic? I don't know, and I'm not in a position to pass judgment, but I don't deny the potential for such things. These people, The People, were an incredible race who did things no one in my time could explain. Crossing thousands of miles of the vast Pacific Ocean long before the advent of latitude and longitude to a tiny island in that vast sea was certainly one. Even in my time, that was a feat that mystified historians.

The Amasiala family consisted of thirty-eight men, women, and children, and they were an ambitious lot. They also brought tools that surprised me. They had axes and knives made of steel, along with spearheads and bush knives. There would be no metals in the ground here, so when those rusted out, they'd be back to wooden tools and weapons, like my spear. Still, there was an abundance of hardwoods in the forest, like the ironwood my spear was made from.

Masotua functioned like a general in his distribution of assignments. Three men were assigned to cut poles per my instructions for the foundations of the fales, three others were to cut the cross beams for the roof, and three more worked at cutting thatching. Still, others were put to work braiding rope to use for the binding. By nightfall, one fale was finished, and two others had the frames completed. Masotua and his wife and two kids spent the night in the finished hut; the rest retreated to the shelter of their boat. I joined those on the boat.

The next day, four homes had been finished, and three more framed out, so things were moving along nicely. Two days later, enough huts had been built for everyone to occupy, and still, Masotua kept building more. The reason, I was told, was that another boatload of family members would be arriving in a few months, and he wanted homes ready for them when they made landfall. Falelima was going to be a full village. Not what you'd call a large one, as it would be later, but sufficiently filled for the time.

It wasn't just the Amasiala family that would be bringing another boatload; there was going to be another armada of boats, one from each of the families. When I asked Masotua if there would be more after that, he just shrugged. My job, however, and those who came with me, was merely to help these first arrivals get established, so we wouldn't be here for any future arrivals. They'd be on their own by then.

Masotua had sent teams of men and women to clear a spot in the jungle for the crops they needed to plant. In a climate like that, everything grows. Poke a stick in the ground, and it will grow. Three days later, a budding taro and ta'amu patch was underway, and inland a bit further, banana trees were planted. Further still, without the need to make a clearing, cocoa, avocado, and some citrus fruit trees were planted.

Fishermen using nets brought in enough fish each day to feed the small village, and divers caught octopus, lobsters, and sea anemones. Hunters brought in fruit bats and a smattering of birds they snared, the feathers of which were used to make an ornate headdress for Masotua, which made him feel quite regal. The meats were cooked in pits, with dry coconut shells used as charcoal. There were indigenous plants that were cooked to provide vegetables since the crops weren't going to be ready to harvest for months.

Sina, Apu, Masana, and Little Sina arrived on the eleventh day aboard what was supposed to be an outrigger canoe. It was crude at best, the worst I'd ever seen. The inner surface where they sat was rough and jagged, the exterior much the same, which would make it harder to paddle, and the rigging was loosely tied in place. In heavy seas, it would likely fall apart. Sina was mad as hell, and none of the others dared to talk to her when she was in that state of mind.

"Petaio is a very bad man," she blurted. "Bad and stupid. And lazy. He will do no good here."

"He thinks he's the king of all The People just because he's building his village in Tufutafoe," Little Sina added. "He actually demanded that Masana stay and marry him. If Sina hadn't been there to step in, he probably would have forced her."

"Why is men's chief now? It should be women's." Sina was firmly ensconced in the way things were done in her time, and I have to admit it had worked out well for them, but the times had changed. I don't know when the world went from maternal systems to paternal, but that wasn't part of our mission.

"There is one woman who is a chief, but only one out of all these families. Things are different in this time."

"Things are stupid in this time," Sina shot back. "Petaio is stupid and lazy. He is not a good chief for his family. And his family is stupid and lazy, too. They build no fale, just take fale we make, and tell us to make more. I wanted to kill him."

"You didn't, did you?"

"I wanted to," she mumbled. "Maybe I will before I go back to my time of good people and chiefs who are women, as it should be."

Trying to brighten her mood, I pointed to the village created by Masotua and the family Amasiala. "This family has done well, even with a man as chief. He has proven to be a good leader for his people. They have built all the fales they need, and are building more for those who will come later."

"There will be more stupid people? The gods are fools, Salini. These villages should be for smart people."

"If you recall, Sina, these villages were captured by the Aunu'u even though they had women as chiefs, which is why we were called to aid them last time. If they were so smart, why did they get captured?"

"They grew lazy," Masana cut in. "They weren't watching or training for an invasion and were taken by surprise. The Aunu'u took advantage of their stupidity and laziness."

"Only in my time are people smart," Sina smirked.

"There are smart people and stupid people in every time, Sina. In my time, there are very many smart people, and very many stupid and lazy ones. It is the way things are in all times."

"Not in my time," she argued. "All people are smart in my time."

"Then you are fortunate to live in such a time," I conceded, not wanting to argue with her while she was steaming over Petaio. In reality, though, I actually wondered if she was right. When I was in her time, learning to be a warrior and fighting with the other women in the caste of warriors, all of them seemed well disciplined and dedicated to their function. And those of the other castes had seemed equally focused on what they did. The villages I knew back then were well-oiled machines. If there were some who were lazy or failed to do their part, I never saw it.

"Aieeeee!" A cry rang out just behind us, where one of the fishermen had just come ashore. He was looking at the outrigger canoe Petaio had made, his nose turned up in disgust.

"You must not go out," he said, pointing at the ocean, "in this. I fix for you." And over the course of the next several days, he did. Our outrigger, when he was done with it, was not only sturdy but also sleek and comfortable, the nicest one I'd ever seen.

"Now you can go. Is it okay for goddess? And queen too?"

"It's wonderful, Akime," Masana said, praising his work. "Thank you so much."

Even with his dark skin, the blush that covered his features was obvious. Another obvious feature was that he wasn't blushing over the appraisal of his work, but by the mere presence of Masana, who was standing next to me. It was pretty clear his work on that canoe wasn't meant for the goddesses who would be using it, but for the queen, Masana.

And use it we did, the very next morning. The Amasiala group, under the able leadership of Masotua, had a totally functioning village up and running, and needed no more help from us, so it was time to head out on the waves to find the next would-be village. Akime was on the beach to watch us off, remaining there until we rounded a bend in the shoreline and could see him no more.

Chapter 6

For the next two months, we traveled the coastline to find where the different families had settled and to help them adjust to life in the tropics. Some of them were doing surprisingly well even before we arrived. They'd built homes fashioned after the ones they'd seen in Falelima before moving to their own chosen spots, and were no longer living on their boats. Some of the homes needed some adjustments to give them more strength against the storms that we knew would come, but that wasn't a major issue. Their crops were planted in cleared areas and were well on their way to becoming good plantations.

There were others who were still living on their boats, waiting for us without taking any initiative on their own. They'd planted no crops, built no homes, cookhouses, or outhouses. In short, they'd spent the time doing nothing. A few had gone hunting for fruit bats, and fishermen continued their craft from the side of their boat, which was the main course in the meals they ate. Divers occasionally went after lobsters and octopuses, but the diet was limited by the lack of vegetables and fruits.

Masana was the expert on edible native plants and spent her time training them on how to acquire and cook them while Apu and Little Sina taught them the building of homes and cookhouses. The outhouses were built on stilts over the ocean in those areas where the jungle was too thick and impenetrable to dig a hole, and the fishermen soon learned that directly under the outhouses was a good place to catch the fish that were attracted to human waste.

We spent the promised ten days in those villages that hadn't done much on their own, the same in more productive villages, and after a couple of months, we thought we were done. It was a premature conjecture, as we found out when a messenger on an outrigger canoe found us, having traveled all the way from Falelima nearly to the far end of the island. He was begging for help over a problem that had developed between Falelima and the self-proclaimed king of Tufutafoe.

It was worse than we thought, even after hearing the message from Akime, the fisherman who'd repaired our canoe and volunteered to seek us out in the hour of need. Petaio had sent his warriors to Falelima and captured several of the men there, whom he took as slaves to build his village the way they'd built theirs. Manual labor of that sort, he claimed, was beneath those of the king's village and should be done by those under his command in other villages.

When we arrived back at Falelima, it was no longer the bustling village we had left two months before. Of the thirty-eight souls who had been there, only a dozen or so remained. Men, women, and even children had been taken to Tufutafoe as slaves to do the work Petaio and his villagers wouldn't do. They were forced to harvest vegetation, plant crops, cook and clean, build homes, outhouses, and cookhouses for the fool who thought himself above all others.

"I will kill this evil man," Sina exclaimed in bitter anger. "Was it not I who said he was no good?"

"Yes, you said that," I replied, "but we must try to correct this problem without bloodshed if it can be done. There are not enough people yet to allow a war between the villages to decimate their numbers."

"Even if it is only Petaio, it is too many people," she fired back. She was ready for a war, even to start one if needed, but I was hoping that Petaio could be forced to abandon his claims to kingship and live in peace with the other villages. About two hours before dawn, we set sail for Tufutafoe, wanting to arrive before Petaio and his villagers were awake.

The first grey lights of dawn were in the eastern sky when we pulled our canoe onto the malae at Tufutafoe, with Akime behind us in his smaller craft. He had cut a limb for a hastily made war club, intending to help in any way he could. There was no activity in the village, but there were dozens of people sleeping in the open on the sandy beach, surrounded by Petaio's guards, who were likewise asleep.

I had wanted Little Sina to stay in Falelima, not wanting her around if a conflict started, but she had insisted on coming along. Masana backed her up, reminding me that she, too, had been but a child, even younger than Little Sina, when she first went to war and had done her part well. "She is the daughter of a great warrior, Salini. Why not let her take her rightful place with her mother? Even in battle if it must be so."

So, we were all there: me, Sina, Little Sina, Apu, Masana, and Akime. Not much of an army, that was certain. Two trained warriors, one who had seen combat but lacked serious training, a child and her father, and a young man who seemed taken by Masana but knew nothing of combat. His weapon was the limb of a tree that may or may not survive the first blow. There was no way of knowing the strength of Petaio's forces, but they certainly outnumbered us by staggering numbers.

We exited the canoes quietly and encircled the sleeping guards on the malae. Then, Sina let out her piercing war cry before I thought we were ready. That brought the village to life in an instant. The guards on the malae jumped up to find themselves surrounded by armed invaders who had removed their weapons, and wisely chose to do nothing that might get them killed.

"Petaio!" Sina screamed at the top of her lungs, "we have come to take away your king's crown and free the people of Falelima from your grasp. If you would fight, come now and face me, Sina, warrior chief of Tufutafoe, defender of the forest. If you send warriors instead, your village will die with you, and you will be shown as the coward you are."

Well, there it was, a declaration of the very war I'd wanted to avoid, but in truth, it had seemed to be one that couldn't be avoided. Petaio rose from his mats, grabbed a spear, and came to the front of the fale he had taken for himself, raised the spear in a menacing fashion, and bellowed out his reply.

"Do you think I fear you, Sina, or the puny force you bring against me? No, I do not fear you, for I have many who will strike you down and let your blood turn our malae red."

A young man stepped forward from the shadowy recesses of a smaller fale, bearing a spear. He turned to Petaio and called out his own declaration. "If there is to be a fight, my spear will be with the spear of Salini," he hollered, then made his way to where we were. He took up a stance next to me, and that was when I recognized him as Masito, the man I'd injured in a short skirmish months before. The scars on his back could still be seen, though they had healed well, as I'd hoped.

"Why do you now stand with me against your chief, Masito, when you wanted me dead before?"

"Because you do what is right, and Petaio does that which should not be done. You are a true warrior, not only in your might with the spear but in your heart, in which only good is found. I am yours to command, warrior goddess."

"You are welcome, Masito," I told him, then turned back to Petaio and those with him. "If any other wishes to join us, they will not be harmed. But if you stand with Petaio, whether because you fear him or think him a king, know that he will not live to see the sun rise this morning if he forces a fight upon us. That is my word, and it is a true word."

“And this is my word,” Sina threw in. “Petaio must die, for he is no king and has done what no true king would do. For this, he must pay, and those who stand with him. This word, also, is true.”

Maybe what she said was true enough, but it wasn’t exactly what I would have wanted at this delicate moment. My goal was to avoid a fight, not provoke one, especially when my daughter, who’d never faced more than schoolgirl pranks in her life, was standing nearby armed with a crude spear she’d never actually used.

“This also is my word,” I shouted. “This is not a time for war. There are few enough people who have arrived on these shores, so why make them fewer? Let us take the people you have captured, admit that you are no king, and have peace with the other villages. Do this and you will live, Petaio. Do not do this, and you will surely die. You and your village, and there will be no king among those who live.”

“I do not fear your words, Salini, for you are no goddess,” he shouted back, waving his spear recklessly over his head and stomping his feet in passionate anger. It was then that I realized the lust for power had driven him insane, which meant he couldn’t back down. “You have four women and three boys against my many. We will see who lives.”

I turned to Sina and lowered my voice. “He’s going to make us fight. Protect Masana and Little Sina. I will go for Petaio.”

“It will be done. No harm will come to them so long as my arm can strike.”

“You are right, Petaio, I am no goddess,” I shouted back at him. “Nor did I ever claim to be. But I am a warrior, trained by the greatest of all warriors who stands with me now, and have seen many men better armed and trained than yours stain the blade of my spear. It is that spear that will be colored by your blood if you force a fight upon us. So, will you destroy your village in a needless fight, or will you submit? It is your word, spoken wisely, that must be spoken now."

"My word is that you shall die," he screeched in a high-pitched tone, and in that instant, he ordered his warriors to take us. If they had come in a united front, as true warriors would have done, it would have been a disaster for us, but the bulk of his so-called warriors hesitated, not sure what to do. I don't think they wanted a fight, but were afraid to face Petaio if they didn't obey him, so they hung back, hoping the bolder ones who lunged forward would win the day.

Out of the forces Petaio had ordered against us, no more than five or six came forward immediately, and they were scattered and disorganized. Some tried to flank us on the right with their makeshift weapons, where I was, while others crossed to the left where Sina stood ready. I pushed Little Sina behind me and ordered Masana to go behind Sina as the charging warriors came at us, telling them both to stay back far enough to give our spears a clear path. And then the fray was on.

My war cry rent the air with Sina's as the sound of clashing weapons broke through our cries. Then there was another high-pitched war cry, that of Masana, and then still another, issued by my daughter who knew nothing of such things but joined in simply because she got caught up in the spirit of things. Sina was fighting like the goddess the legends claimed her to be, dropping two of her attackers before having to dodge any blows herself. Masana, behind her, dropped another one while I took on two on my side of the fracas. They went down before I even realized I had struck.

It was always that way with me. When a fight was forced on me, I seemed to lose my grasp of reality and fought with an abandon based on training, not conscious thought. I never knew how many I had killed or injured until the battle was over. But I knew that my daughter was behind me, and that no enemy could be allowed near her. I guess I failed in that because when the fighting was done, she was bawling like a baby, kneeling next to the body of a man she had killed.

Petaio, in the meantime, was screaming at those who had failed to attack as ordered, prodding them on like a maniac, but by then they had seen how those who obeyed him fared and wanted none of it. Petaio made his worst mistake then, striking down one of his own men with his spear. Enraged by that, some of his men turned on him, backing him once again into his fale. It offered no protection for him, but the would-be warriors who were after him hesitated to go inside, where the use of weapons would be limited by the size of the dwelling.

I didn't. Running to his fale, I burst inside and turned to face him, my spear up and ready. His eyes were frightened, but with a fire that I knew wouldn't let him concede. His spear came up as he charged, but he never got to bring it down. My spear's blade opened up his unprotected mid-section, and he fell to his knees, his guts spilling over the mats. His wives ran from the fale in terror as his spear fell to the mats at his side, his eyes glaring at me.

"You can't kill me," he blurted, his words filled with pain. "I am the King of Mutasi."

"You are not the king of anything," I retorted fiercely, "which is why I can kill you. I have already done so, for you will bleed to death in minutes, but I have no minutes to spare." With a wide sweep of my spear, I swung at his neck, where the teeth virtually crushed his spinal cord, killing him instantly. He dropped on the mats, flinched once, then lay still.

It was not yet the time of the rising sun.

Little Sina was on her feet, having watched it all, and came running to me with a bear hug I desperately needed. "My god, mom, I don't believe it. It was insane."

"War always is," I said, running my free hand through her hair. "Any war, any time, any weapon."

Sina came over to us with a huge grin on her face. "I tell you she great warrior, yes? But I wanted to kill Petaio. I will not forget this injustice you have done to me, Salini. Maybe, though, we can still be friends?"

"Always, Sina, war chief of Tufutafoe and defender of the forest. Friends forever, through all time."

"That's when the captives from Falelima surprised me when they took up the weapons of those who had fallen and started to charge what was left of the Tufutafoe men. I had to jump in their way and cry out in the fiercest voice I could manage, which was pretty fierce at that moment. 'If you would press this fight further, you must first fight me,' I challenged them. 'There has been enough killing this day. Lay down your weapons and return to your village in peace.'

Then I turned on those of Tufutafoe. 'There lies the man who claimed to be your king. Deal with his body as you wish, but let no one claim to be king again, or we will return and do likewise to him. If you would live, let there be peace among the villages from this day forward. That is my word, and my warning to you. Let it be so.'

'We did not think him our king, but served him out of fear,' one man called out, laying his spear gently on the ground before him. 'We will not fear another again. That is my word. I am Fenosi of Tufutafoe, of the family Sa'atofu.'

'Your word is good to my ears, Fenosi. It is in my mind that you would make a good chief for your family if they will so honor you. If it be so, do not think yourself more than those who follow you. That is the path to destruction, as you have seen this day.'

'He will be our chief,' a woman called out, 'but if he does not do as a chief should, we will strike him down as you have done to

Petaio. There will be peace now, in our village, and with the other villages.'

'It is good,' I nodded. 'Now we will take the people of Falelima back to their village. You must clean up the carnage of this battle on your own, for it was you who made it so.'

'Let it not be so,' one of the women from Falelima cried out. 'If there is to be peace among us, let it begin with our help in cleaning up this battle. When that is done, we will return to our homes, knowing that we did not abandon our friends to such a deed without help.'

And so it was. Those from Falelima helped to clean up the blood and guts, drag away the bodies, and even dig holes for their burial. I was relieved to see that they didn't require the severing of their heads as was done in Sina's time, for that was something Little Sina didn't need to see. Sina, for her part, thought these new arrivals were cruel for not doing so, as the fallen could find new gods in their next lives. To their credit, though, the bodies, once laid in graves, were crushed by huge rocks to accomplish what the severing of heads did in Sina's time.

The body of Petaio wasn't buried. It was dragged out into the ocean and dumped for the fish to feast on. It was, I was told, the ultimate insult that could be inflicted on a body.

'Our mission is complete,' Sina said, taking me aside so no one else would hear that we were going to leave.

'It would seem to be so,' I agreed, hoping beyond hope it was over. This had not been the peaceful mission the Little People had promised, but we had done what was asked of us, and all else that needed to be done. 'I will have Masana contact the Little People that they may return us to our own times. It has been good to raise my spear with you again, Sina. And it is always good to stand at your side, whether in times of war or peace. It is my hope that you will remember me with fondness, my friend of friends.'

‘And you will think of me from time to time in a similar manner, Salini, adopted daughter of Tufutafoe and warrior from Siata? We are sisters in peace and war. Let it always be so.’

‘It will always be so,’ I promised, and meant it with all my heart.”

Chapter 7

The response from the Little People, when Masana contacted them in the way they had instructed, was given in person. The damn creature informed us that our mission wasn't complete after all, despite our expectations. The spokesperson, or thing, or it, was obviously somewhat embarrassed by the message it delivered. It confirmed that we had completed what had been asked of us for this mission, but added that a problem had arisen that made it necessary for us to stay longer.

"Why is this so?" I demanded. "We have done what you asked of us. Even more, for you promised no violence on this mission. Why, now, is our work not done?"

"An enemy approaches," it said with a shake of that head over the scrawny neck and a blink of those overly big black eyes. "A fierce enemy who comes to steal the islands of The People and enslave them. This cannot be allowed."

"Who are these people?" Masana asked.

"They are of the black-skinned people that left Mutasi while it was yet young. Now they wish to return, and have been told that these islands are more desirable than those they should have gone to. This we did not expect."

"You speak of those who will be the Micronesians?" I insisted. "Or the Melanesians?"

"It would seem so," it replied with another nod and blink of its eyes. "They were supposed to stop at islands far to the south, but did not do so, and now they are coming here to take these islands. If they do so, what you have done will be for nothing."

"How many of them are there?"

"In your way of counting, maybe a hundred. Or maybe more. It is not certain."

"What weapons do they carry?" Sina asked. I think she relished the thought of another war so close on the heels of the one we'd just won so easily.

"They have heavy clubs, and sticks stretched tight by a string that fires sticks through the air. These are dangerous weapons. Do not take them lightly."

"I do not fear sticks," Sina blurted proudly, "but they will learn to fear my spear."

"They are bows and arrows," I interjected into the discussion, "and they can kill at long distance where a spear only kills up close. They are, indeed, to be feared. Are the points of the sticks made with iron or steel?"

"They are sharpened stones. Very sharp, to penetrate deep into the body."

Probably flint, then, or stones hard enough to be chipped into a sharp edge and tip. I'd dug up such arrowheads in my archaeological digs, and knew just how effective they could be in the hands of a skilled opponent. Flint is formed when lava falls on snow, ice, or cold water and cools rapidly, but none of those conditions existed here. The lava here had cooled slowly, releasing gas fissures that made it weak and brittle.

"How long before the invaders reach here?" I asked.

"A cycle of the moon, as you measure it. Maybe more, maybe less."

"That's not much time to prepare for a full-scale invasion."

It nodded again. "But it must be done, for they will kill The People with hard labor and little food, which they will keep for themselves."

“What can you do to help us in our preparations? Can you turn them back to where they should have gone?”

It shook its head. “We can do nothing but warn you of the danger that lies ahead. Now you have been warned.”

“Can you at least take my daughter and Apu back to their time? They are not warriors.”

“It cannot be done. All must go, or none. If all were to go, The People would die.”

“But that would change the events of time, which can’t be done,” I argued. “That’s what you told us.”

“Then destroy the invaders, or send them back to where they belong. Then the events of time are secure.”

“I don’t believe you. I think you lie, making us think you know the events that transpire over time, but you do not.”

“What is it you think we are?” it asked.

“You are gods,” Sina barked at it. “That is what we thought you to be, and what you claimed. Is it not so, now that we are in need of gods?”

“We are not gods,” it said with another deliberate shake of its head and blink of those giant black eyes. “We know more than you for we can see more, but we are creatures of this place, like you. We are not as strong and courageous as you, and cannot fight as you can. We can only tell you of dangers that approach. The rest is for you to do.”

“Tell us what their boats are like.”

“They have a wooden frame, and reeds sealed by resin for the skin. They are fragile, but do not sink and are capable of long journeys. They will not fail them.”

With that, the monster simply levitated a foot or so off the ground and floated away in the damnable way they do, presumably to its ship which was most likely nearby. I had the distinct impression they couldn't walk on solid ground. Other than the warnings they gave us, they were as useless as tits on a boar hog, and I was glad that I'd defied them my first time into the past. And so had Apu. No, they weren't gods and never had been.

"Well, we're stuck again," I announced to those who hadn't been able to hear the telepathic messages from the creature. "And as usual, we're facing another war. And this time it's a big one."

"Yes," Sina added. "Bigger than any we have fought before. Are these air sticks truly to be feared?" It seemed obvious that this time she'd been able to hear the telepathic message.

"Yes, they are to be feared. With them, they could kill us without coming on land. We must prepare differently this time if we are to win this fight."

"Who's coming, mom?"

"People with black skin who ran from our lands long before my time," Sina explained.

"Why did they leave?" I asked.

"I think we made them go," she explained. "It was long before my time, so I know little, but I was told that we did not like them and made them go away. It is said they were dirty, and mean, and did things to hurt Mother Earth, so we made them go away. The legend says it was a huge battle."

"Where did they go?"

"I know not of this. To land of their own, I would say. Where that is, I know not."

"So now they're coming back for vengeance, probably from a place called Africa in my time. They want to even the score for what was done to them in very ancient times."

"Maybe, maybe not. I know not of these things except what has been sung by my people.

"It matters not who they are, why they are coming, or where from, it matters only that they are coming, and we must be ready to defend ourselves. We've got much to do, and little time to do it. They'll probably hit Falelima first, since that seems to be where everybody is coming to. So we begin our work there, and we'll need all the help we can get."

Sina looked depressed. "They are so many. How can we fight so many?"

"We'll have to divide them into smaller groups," I said, "or cut down their numbers before they arrive. There might be a way to do it, but it will require the cooperation of all the chiefs. We need to get them all to Falelima, with their boats and as many men as they can bring. Somebody will have to go to the outlying villages to bring the word."

"I go," Apu piped up. "I take canoe and go fast, bring all chiefs and all men. It will be many, for this threatens all."

"Thank you, Apu. But be careful. Some of those villages weren't as agreeable as we'd hoped they would be, so maybe they won't be agreeable on this, either."

"I tell them they help or enemy kill them."

"I chuckled. 'Not very subtle, but it should work. We'll be waiting for you at Falelima.'"

Apu ran off to arrange for a small fishing paopao, or outrigger canoe, while we went back to explain the pending threat to the villagers. Some were reluctant to believe us; others quickly offered to help in any way necessary. Fenosi stepped forward, pounding his chest with a fist. "Tufutafoe will not let you down. Tell us what to do, warrior, and it will be done."

"Thank you, Fenosi. We need to get many strong men to Falelima to help us prepare our defenses. Take them there on your large boat, for we may need it also."

"We will do as you say," he nodded emphatically.

"Do so quickly, for there is little time to waste."

"We go now, Salini. It will be good to raise our weapons at your side, rather than against you."

"Yes, it will be good. And we will show you how to make better weapons to use in this fight."

Fenosi ran off to gather his people for a meeting, which gave Sina the chance to ask further about my plans. "For what do we need the large boats, Salini?"

"The boats of The People are made of hardwood, and are strong. The boats of the enemy are covered with reeds and sealed with a resin which burns quickly. If we can get close enough to throw torches at them, their boats will burn very fast, and they will drown, or burn with their boats."

"But what about their flying sticks? Won't they throw them at us before we get close enough to throw torches?"

"Yes, but if we build a wall around our boats strong enough that the sticks can't go through, we will be safe. And if we can't burn their boats we can ram them, for ours would not be damaged, but theirs would likely be destroyed."

"Can such a thing be done?"

"It has been done at other times, yes. And I will teach The People to make flying sticks which can be sent with fire on them. Then we won't have to get as close to their boats, and will be safer."

She got that childish grin that covered her features when she was excited. "This will be a fight like none we have seen before, and the victory will be sung for all time."

"If we win, that is. There will still be some of the enemy who will reach our shores, so we must have strong defenses built on land as well. There is much that must be done. What oil was used by The People in the lamps they brought?"

"This I do not know, but I will ask."

"Whatever it is, we'll need a lot of it. Find out how much they have, or where it came from."

"I will learn of it, Salini. Ayeee, it will be good to be in battle with you again."

I didn't share her enthusiasm over another battle. Little Sina was with me, and Apu, neither of whom had ever been in a serious fight before, and I didn't want them involved in the one ahead. It promised to be a bad one, and no one could be considered safe. No one. Not them, not Sina, and not me. Then there was Masana. She'd seen battle and done her share of fighting, but she wasn't born a warrior. She was a princess, or queen, as some had designated her. She was born to rule, to send the troops out to do her fighting for her, not to wield a spear against an enemy with archers.

This was going to be a war the likes of which The People had never seen or had experience with. There would have to be a sea engagement, armada against armada, something they would have to learn, but would have little time to train for. It was going to require a lot of preparation and tactical planning. If we were going to win, we would have to do so by wit and the element of surprise, not through a face-to-face, one-on-one confrontation.

The way it looked at the moment, it was going to be up to me to organize the planning, preparations, and training we'd be needing. The problem was, I had no experience with it either. But I did have one advantage. I'd read about battles in history that might be similar, that would give us a bit of an edge when it came to preparations and implementation. They were battles that might suggest tactics and strategies that could possibly help us as we approached the challenge that lay ahead.

The archers were the big issue. We had to find a way to protect ourselves from them, to take away the advantage they gave the enemy. And then we had to have better weapons than they did when the fighting got to close quarters. They had stone-headed clubs; The People had axes of iron, but it was still club against axe. Their clubs had wooden handles; we had wooden spears and javelins. Some on either side might break, or they might not. That kind of fight, face-to-face with a determined enemy, would mean a lot of casualties on both sides, probably in equal numbers, and they greatly outnumbered us.

We had to find a way to make the casualties so one-sided, so devastating to the enemy forces that they would either be wiped out or flee. Either way, we would win. Failing that, we would most certainly lose."

Chapter 8

Fenosi turned out to be a surprisingly capable leader. His family worked hard, but not because he forced them. He had a way of instilling confidence in them about our coming victory and gave them the incentive to work harder to make it happen. The other families helped also, since no one wanted to die, but none with the intensity of the Sa'atofu family. Fenosi, though now a chief, shed as much sweat laboring in the various facets of our preparations as anyone else. He was a man to admire.

Within a week, four of the large family boats had returned to Falelima and were rigged with a wooden wall around the exterior to protect us from enemy archers and boarders. The walls were neither strong nor perfect, having been made with crude tools. There were openings an arrow could penetrate, but they were narrow and could only be hit by a skilled archer. We hoped we wouldn't have to deal with many of those. My plan was to make bows and arrows of our own, with flame-tipped arrows. If luck held, we wouldn't have to actually ram the reed boats, just get close enough to shoot flaming arrows or throw torches.

I'd never made a bow in my life, but assumed it wouldn't be that hard. Wrong. There were so many things that had to be taken into account when trying to make an effective bow, not to mention an arrow, that it took me the better part of three days to get just one that pretty much worked. The upper and lower parts of the bow had to be equally balanced in terms of girth, length, and strength, and there had to be some sort of rest at the middle for the arrow to slide across without ripping your finger off. And the bow had to be strong in order to launch an arrow the distance we would need, especially an arrow weighted down with some sort of burning resin.

The arrow had to be reasonably straight, of solid wood without knots, and needed flight guides on the back to keep it going in the direction intended. That meant feathers, which were in abundance on the island, but glue to hold them in place was in short supply. We finally had to settle for the resin of some rubber trees we found, which was a long way from glue but better than nothing. After the feathers were rubber-mounted into place, the three of them were tied with a tiny string of bird sinew.

Spears were another major issue. With the tools available to us, we didn't have time to harvest, season, and carve intricate tupas like those Sina and I used, so we went with a cheaper substitute. Most of the ones we made for The People had no teeth like ours, or just one or two on each side. To compensate for that, we made longer blades that were moderately straight. Because the wood wasn't properly seasoned, though, the blades weren't as strong as ours, which made them less stable and, perhaps, not as deadly. They might tend to crack along the edges, so we trained the fighters to use stabbing motions when possible, not slicing or cutting.

Javelins were the easiest thing to make, and we made hundreds of them, but they were pretty much a one-use weapon. Once it was thrown, it either did its job or fell into the hands of the enemy who could then use it against us. The short spear I'd first used in my training was deadly, but too intricate to be manufactured in such a short time with limited tools. It also had the handicap of requiring even closer combat.

Upon their arrival, The People had brought lamps made of clay, holding some kind of burning oil. That, we discovered, was whale oil harvested far to the south where the water was colder and whales were abundant. That wasn't the case in the warm waters surrounding Samoa, or what would become Samoa. But we needed something that would burn and ignite quickly like oil, which was a problem. A member of the Kusani family came up with a solution that seemed to work.

It wasn't perfect, but it beat anything else we could come up with. For our land defenses, we would dig a long trench across the malae, the only place they could attack from, with a bed of coconut shell briquettes burning in the bottom. Over that would be laid a false ceiling covered by dried reeds that would ignite immediately upon falling on the charcoal. When the enemy stepped on what they thought was just a layer of dried seaweed, they would fall into the pit and be burned.

Our defenses for a land engagement were about as complete as we could make them, but only at Falelima, where we expected them to land. If they hit anywhere else, we would be defenseless, so our primary goal was to take the attack to them on the open sea. A sea offensive was as foreign to The People as it was to me, though, so it took a lot of deep thinking and planning.

It was one thing to fortify our boats against archers, but that also made them heavier and less navigable than the lighter boats of the enemy. As long as we could get close enough to launch our arrows and torches with accuracy, it worked in our favor. The problem was that it made it more difficult to overtake the enemy fleet if they were to flee to another landing point. We had to reduce their numbers in the ocean. That was the bottom line. This attack probably wouldn't be as easy to foil on the ground as the Aunu'u were on my last mission.

So there I was, strategist and war general, with everyone thinking I knew what I was doing. Except me. Both Sinas, all the chiefs, and even Apu were quick to take on any task I gave them, trusting that it was part of something brilliant. Masana was my personal cheerleader, going around to bolster courage and trust in my leadership without being asked.

All the while, I had no idea what I was doing. The closest thing to experience I had was reading about an early Greek sea battle and perusing a book about WWII battles in the Pacific. Neither gave me a clear map of what to do in this case. I was relying on a lot of things that may or may not be true, and if I was wrong, a lot of innocent blood would be spilled. I assumed that the word "resin" referred to some flammable substance. If that was the case, igniting it could be a game-winner for us. But if it didn't burn? What then?

Ram their boats? If I were at home with my computer, I could have done some research on reed boats, but not here. Surely the reeds would burn, but how quickly would they ignite? And just how strong of a structure did such a boat have? What if we rammed it and it turned out to be stronger than ours? I was putting way too much emphasis on assumptions, and the only place to find out if I was right would be on the field of battle. Or high seas.

Sina had taken over the task of teaching The People to make and use spears, knives, and clubs. Part of that training involved the art of deception, ways to attract the enemy to where you want them as we'd done with the Aunu'u on our last excursion. Masana was the expert on hiding weapons where they could be retrieved quickly when needed, and Little Sina had found her niche in helping children and mothers with small children find suitable places to retreat when the time came. Women unencumbered by children were expected to fight with the men.

Training the archers fell to me, despite the fact that I'd shot a child's bow a couple of times in Girl Scouts and nothing since. The problem of trying to teach what I didn't know really showed up, however, in the making of the bows and fashioning of the arrows. The first dozen or so bows had no power at all, capable of launching an arrow no more than twenty feet, and that without any kind of accuracy because the arrows weren't straight enough and the feathers we used for fletching kept falling off.

It took a week before we managed to come up with a bow that launched an arrow a hundred feet, and once we figured out the problem, we soon had several that were good for up to two hundred feet. The bowstring was another issue. The thin chord made from coconut husks was strong enough, but it would unravel after a half dozen shots or so. It was an enterprising young girl who came up with a way of interlocking the fibers in the weaving process that saved the day for us.

Within two weeks, maybe slightly more, we had about three dozen solid bows and several hundred arrows that could be shot with relative accuracy, enough to hit a large boat at least, two hundred feet away. We weren't going to win any awards in a modern tournament, or even an ancient one for that matter, but we hoped it was good enough for what we needed.

I picked out ten of those who showed the most promise with a bow, two of which were girls in their upper teens. My plan was to put two boats to sea with five archers in each one. Our goal was to burn their boats with flaming arrows. If we stopped them far enough out, they'd drown when their boat was destroyed unless they got picked up by another one. Which might, hopefully, overload that one and make it easier for us to overtake, and burn, as well.

I figured I'd need at least ten strong men on the paddles for each boat but selected fifteen to cover any losses we might incur. Our routine became a never-ending cycle of training. I'd take the boats out each morning for a practice run against another of our boats, then we'd come back and everybody trained on spears, knives, and clubs. Even our sea engagement personnel could wind up in close-combat fighting, and I wanted them prepared just in case. Everybody who wasn't with child or caring for young children trained on weapons.

Our first open water training exercise was a disaster. The target boat was taken out well beyond the reef and its crew transferred to another boat as a precaution. Then we launched our attack boat, paddling furiously to make contact and get within bow range before the target (ours was immobile, but we recognized that a real target wouldn't be) could evade us. That's when we discovered that it took a lot of hard rowing to get our heavy boat moving, and the same or more to slow, turn, or stop it.

The collision that occurred between our assault boat and the target left both in need of repairs. Nothing serious, but nothing to ignore either. It did, however, demonstrate how effective a ramming operation could be, especially if the enemy vessel was built on a lighter superstructure than ours. In my mind, reed boats would most assuredly be built on light superstructures, but I didn't really know anything about them from personal experience. That was another assumption I hoped wouldn't come back to haunt us.

Another thing we learned in that first training mission was that arrows fly differently in the wind over water than they do in the relatively protected air over land. That started a series of lessons on windage, another subject I knew nothing about and had to learn on the fly. "If you aim at a boat but your arrow lands somewhere else, adjust your aim accordingly," was about the most sophisticated way I could explain it. I think the archers were way ahead of me on that one, though, and were doing it even before I tried explaining it. Maybe they didn't have experience with bows and arrows, but they weren't short on common sense.

Thank God, because too many times I was.

I began to think more and more about using ramming as a primary offensive maneuver based on that first test run, but it wasn't without problems, either. First, it virtually ensured close combat conditions, because it provided a good opportunity for the enemy to jump aboard our craft at the time of impact. In fact, the force of the collision might even make some of them fall onto our craft. In that case, my warriors would have to switch quickly from bows to spears or clubs, or I'd have to bring extra fighters to provide security in that situation. Sort of like marines in naval forces. But I didn't have an endless supply of warriors, and didn't want to deplete my land forces.

Sina just shook her head at my dilemma. To her, it was all very simple. "Three warriors with spears not let bad mans come on boat," she declared. And yes, there was some wisdom there, but that meant three more people on each assault craft that had to be shielded from enemy arrows while fighting in an exposed position. And three less on the ground for each boat we took out.

"What you need is Constantina wire like they use in prisons," Little Sina observed. "Or something like it, anyway." And that was it. I couldn't believe such a simple solution had evaded me, and everyone else for that matter, but there it was. Sharp stakes—javelins—firmly attached to the front of each boat, facing forward so as to impale anyone trying to cross over. Close enough to each other that an enemy couldn't slide between them, and strong enough that they couldn't be pushed aside or broken off. That started another production line for an entirely different kind of weapon. They didn't have to be pretty, and they didn't all have to look alike. They just had to be strong. And sharp. Very sharp.

A little over two weeks into our preparations, we were starting to feel more confident in our ability to not only survive the coming attack but win it. There was an optimism that permeated everything, which could be a good thing, or bad. As long as it didn't result in a feeling of false security, a letting down of our guard, it was probably a good thing. We had to meet the threat with brutal force, solid training, and a determination to win at all costs. Nothing less could be acceptable.

Sina took me aside, pulling me from the lines of women making bowstrings. "We have done much to prepare for a fight, Salini, and The People have worked hard. They not fight each other, make things we say to make, learn to be warriors. They are much good people. Yes?"

"Yes. They are much good people. I did not think they would be so good, but I am pleased."

“I am also pleased, but The People do not know this thing.”

“What thing?”

“That we is pleased, Salini. It is in my mind that we must have a fiafiaga.”

It struck me like a lightning bolt. Yes! A celebration of all that had been done was exactly what we needed, and it should be the grandest party ever. That was the way of The People. I’d been so involved in planning for war that I’d forgotten how these wonderful people lived. How they expressed themselves, how their emotions were channeled into productive activities that benefited the village and each individual in it.

In most societies, planning for a party is a complex operation. Not among The People. All we had to do was make the suggestion and they were all over it. Little Sina, who had been adopted by the mothers and elderly women who did most of the cooking, got immersed in food preparations. Masana helped to organize the music and dancing, Apu got involved in the making of torches for the malae, and Sina worked with the chiefs to have a kava ceremony like none before. It didn’t seem like I was needed for the party preparations, and frankly I felt left out. But I was grateful for it. I’d been too swept up in too many things, so it was nice to let the others do the planning for a change.

That feeling got swept away in a hurry when the fiafiaga got started in a dramatic way. A way nobody planned for, and a way nobody would have wanted.

Chapter 9

The kava ceremony got underway as they always do, and it was a good one. Along with the eleven chiefs, Sina, Masana, Apu, Little Sina, and myself were all given seats of honor. Little Sina was getting a crash course in the Polynesian lifestyle and way of thinking, and she enjoyed every minute of it. I didn't blame her. I was pretty partial to it myself.

Her mouth puckered noticeably with her first taste of kava, just like I had done my first time, but she recovered well and didn't make a scene. Kava has a flavor that isn't easy to describe. It's kind of like dishwater with a sprinkling of metallic flavors. Once you get used to it and know what to expect, it isn't bad, but that first shot will take you by surprise. And the way it's served doesn't exactly follow the norms Little Sina had been raised under.

She was surprised that there weren't individual cups, that everyone drank from the same one. In her upbringing, that would be considered unsanitary, but to her credit, she didn't hesitate when her turn came up. And the fact that the cup is mixed and served by a topless young woman made it even more bizarre, though it added to the ceremonial aspects. All in all, she accepted the whole thing quite well, even stumbling through a rather pathetic acceptance speech. It was rare, though, for someone as young as her to be in a kava ceremony anyway, so no one expected a skilled response.

Apu, who'd come from an environment where men weren't allowed to take part in kava ceremonies, was nervous and had to struggle to deliver an obviously simple speech. Both he and Little Sina forgot to make an offering to the gods. This is done by pouring a few drops over the front of the cup onto the mat as you begin your speech. The speech is expected of all participants, and in formal ceremonies, each can last several minutes. Talking chiefs, and some high chiefs, usually deliver theirs in the respect language.

The feast began when the kava was finished, and I got a kick out of watching Little Sina as the mats of food were brought in and placed before us. She kept licking her lips, trying to get some feeling back in them after the numbing effects of the kava, but it was no use. The numbness would wear off soon enough, but nothing she could do was going to hurry it along. I was curious to see her response to the meal as well since a fiafiaga feast is nothing like the regular meals she'd shared with her new friends.

I never got the chance to see her reaction, or enjoy the feast myself. None of us did, because it suddenly came to an end just as it was about to get started.

They came out of nowhere. At least, no one noticed them until there they were, standing with the ocean swells lapping against their heels, facing us menacingly across the malae. They were black as night, two men, tall and lithe, with cropped hair and muscular torsos. They held what had to be battle axes, for there wouldn't be much else they would be useful for. The heads of the axes were formed from a solid mass of obsidian, sharpened at the cutting edge and blunt on the back.

"Hear me and listen well, people of this village," one of them shouted in our language, his deep voice carrying over all other sounds. "We are not here to kill you unless you make us do so. We are here to kill the fearsome goddesses who have enslaved you."

It was Fenosi who recovered his wits quickly and was the first to respond. "You are mistaken. There are no slaves here."

"Is it so? My eyes see many. My eyes see slaves who have built a village where they had no right to build, forced to do so by goddesses who have no right to live. Goddesses who are called in the legends Sina and Salini."

My hand rested on my spear as if it had a mind of its own, curling about the shaft and lifting it from the mat where it lay next to me. Sina, sitting just a few positions to my right, did likewise, then stood to face the black menace.

"I am Sina!" she bellowed like a call to war. "Whether I be goddess or not, it seems it is I that you seek."

"And the white-haired one," the black spokesman said, pointing his battle axe at me. "It is you we have come to kill, not the people you enslaved."

I started to rise, but Little Sina grabbed my arm. "Don't, Mom. They want a fight."

"So it seems," I said, shrugging off her clutch, "and I don't think they intend to be denied."

"I am Salini," I called to them over the soft sounds of the surf. "Who are you, dark assassins?"

A grin, showing bleached white teeth, pierced the growing shadows as night approached. "We are god killers," he shouted back for everyone to hear, "who have come to free the slaves."

"There are no slaves here," Fenosi called out again.

"My eyes see many who have built a village where they have no right to live."

"And where is it you say we should live?" Fenosi pressed.

"In the stomachs of true men, which will be your destiny."

Cannibals. Nothing new to me and Sina, who had fought them before when women did the fighting. But the reaction we saw on the faces of The People told us they had no concept of such a thing.

"Why do you wish harm upon these people?" I called out again, wondering if I knew the answer already.

"No harm is wished upon them, goddess. We wish them fat, and tender, and juicy."

"And then you kill them."

"We tend our animals well, goddess. They live to bring more young into the tribe, so that we never run out of fresh meat. We kill only as the herd dictates."

"These are not animals, nor do they make a herd. They are people, and you shall not kill them."

"My eyes see animals who cannot fight, who need goddesses to fight for them."

"We can fight!" Fenosi screamed as he charged, spear in hand.

"Fenosi, stop!" I shrieked, but he either didn't hear or paid no heed. He bolted to the one who'd been speaking, but was still two strides out when the other one cut him down with a single stroke of his axe that opened his chest. Fenosi fell, eyes wide with pain, and then they closed as he lay face up on the sand. There was a slight tremble, and then he moved no more.

"The harvest has begun," the man with skin the color of night grinned.

"You shall harvest no more," Sina shouted, then followed that with her piercing war cry as she lunged forward. Not willing to let her face them alone, I sprayed sand behind me as I drove myself forward. There was no time to think or plan our attack. We had seen how sharp and strong their battle axes were, how easily Fenosi had been fatally opened up, so we knew we had to avoid getting hit. That was about it. We knew nothing of their fighting techniques or strategies, the best way to come at them or from what direction. We knew only that they had killed a good man and needed to be stopped before they killed again.

Then again, it seemed that we were the ones they planned to kill anyway, so by charging them we were playing right into their hands. Our plan was to hopefully turn their ambitions against them, but it could obviously go either way.

From the corner of my eye, I saw an axe swing through the air, driving just over Sina's head as she ducked beneath it and came up with her spear already arcing toward her enemy. He jumped back in time to avoid having his stomach sliced open, and in the process lost his footing in the loose sand. He fell, with Sina in hot pursuit. That's where I lost sight of her fight, because mine was on me by then.

The assassin's axe was coming straight for me, so from habit I brought my spear up to parry the blow. It hit like a sledgehammer. The axe was heavier than my spear, the arm swinging it stronger than mine. I fell back three or four steps before I caught myself, then dug in for the next blow. It came at eye level, easy to duck under, and, like Sina had, I came up with my spear in motion. It caught the black demon just under the rib cage, opening up a long wound, but not a serious one.

The black assassin gave me a surprised look, running his fingers gently over the gash in his body, then bringing them to his lips to taste his own blood. It wasn't a look of fear, just one that said he hadn't expected such a quick response to his attack. He didn't back off, though, and swung his axe again, this time with more force than before. I backed off to make it miss, but that increased the distance between us so that my spear couldn't reach him for a counterstrike. He lunged forward and swung again, this time in a downward stroke intended to crush my skull. I managed to sidestep it and bring my spear around to his left side, scratching his skin there, but not deeply.

Again he looked at me with those piercing eyes, obviously my ability to respond to his blows was something he hadn't expected, but there was still no fear in him. No fear, just intense anger. Fury. Hatred. An emotion that could be taken advantage of. Raising my spear over my head, I brought it down in a powerful blow at his head, which he parried to the left, but I used the momentum of that swing to bring my blade back up under his defenses. It cut a line up his side and into his armpit. Not the one holding the axe, but it was a painful wound nonetheless.

I don't know what he screamed at that point. It was no word I understood, but I knew it was his version of a war cry as he charged straight at me, his axe raised over his head. I jumped back, tripped in the sand, and landed intentionally on my back in a carefully choreographed fall. He rushed forward to bring his axe down on my unprotected position, and ran right into the point of my spear, brought up to take him as he charged in.

Blood dripped on me at first, then started gushing. It was a fatal wound, but not one to stop him instantly, and he didn't. I had to roll to the side as his axe came down, more as a fall than a planned attack, which opened up the hole in his stomach even more. He paused to look at it in shock, and I used that pause to drive the spear into him again, this time in the groin. That dropped him in agony, and with him down, I jumped up and swung my spear in a powerful arc that severed the back of his neck, killing him instantly and ending the fight.

Remembering Sina, I turned to find her in a hand-to-hand match with her rival. His axe and her spear were both on the ground, apparently fallen in the struggle, and each trying to prevent the other from being the first to get his or her weapon. I rushed over and drove my spear into the side of the black monster. Taken by surprise, he jumped back and released his grip on her. That allowed her to grab up her own spear and swing it in a wide arc that tore into his back, severing his spinal cord. He went down silently, without so much as a twitch.

For a few minutes, we knelt side by side in the sand, oblivious to the blood and gore, leaning on our spears as we gasped for air.

Then Sina turned to me. “It is not good what you have done, Salini of Siata.”

“What did I do?”

“You helped me to kill this thing, but I wished to kill it myself.”

“Yes, it was a foolish thing to do,” I retorted, “but I did not think. It will not happen again.”

“Aiyeee,” she cried out as if to heaven, lifting her head back to the sky and laughing out loud, “it was good to be in battle again with you, Salini.”

“Let us not do this often, Sina, war chief of Tufutafoe.”

“That would be good,” she chuckled, then jumped to her feet and offered me a hand. “Again you have done that which kept me from dying. I am in your debt.”

“We are friends, Sina. Such friends as we have no debts.” And then we were surrounded by The People, and there was no further chance to speak one to the other.

After a frenzy of congratulations from everyone and serious hugs from both Masana and Little Sina, we managed to bring things to order. The first thing we had to worry about was if there might be more of the enemy about, so men were sent out to search the surrounding area in groups of five, all armed. It didn’t take long before a small paopao, or outrigger canoe capable of carrying no more than two, was found tucked away undergrowth not far from the village. There was no sign of any other craft or enemy forces.

The fact that those two had arrived in such a small boat, however, suggested that the main force couldn’t be far behind. That meant our time for partying was over. Our time to prepare for war was over. The war was upon us. In fact, it had already begun.

Chapter 10

Fenosi was quickly buried, but with all the honors due such a man. It was Little Sina who came up with an idea for the bodies of the two assassins. She suggested that they be hung at the front of our attack boats to show the enemy that we weren't an easy target, that their assassination plot had failed. How such a kind, friendly girl like her came up with such a grisly plan is beyond me, but the villagers liked it and went to work putting it into action.

With any luck, it might have a small impact on the enemy, but I was still worried. Those two men had been more dangerous than any enemy we'd faced before, except for Big Tooth. If they were indicative of what the rest were like, we were going to be in a lot of trouble. This wasn't going to be a fight against drunk Aunu'u guards, it was going to be a fight against strong and skilled opponents. Stronger and more skilled than us, almost certainly.

There was a quick council meeting with the chiefs in which it was decided that no torches or fires would be lit that night, and the attack boats would be ready to launch at first light. The women and children who were not combatants were taken to their place of refuge, joined by Little Sina, which made my heart feel a little lighter. This was going to be a battle she didn't need to have any part in. Lookouts were posted at several sites along the coast to watch for incoming boats, while those who were going to be on the attack boats labored to get the needed supplies and equipment on board.

It was a tense night. The assassins had come on a tiny reed boat that carried no supplies, so they had to have left their fleet not far offshore. It wasn't feasible to assume they'd spent days paddling without rations, so the enemy armada had to be nearby, just beyond eyesight. Would they attack at night? We were ready if they did. As ready as we could be, that is. We needed the advantage of that open sea engagement to reduce their numbers if we were to succeed, but putting our boats out in the darkness would be foolhardy.

Everybody spent the night at their assigned positions. The ground forces surrounded the malae and listened to Sina's urgent commands. Those who manned the boats spent the night on them, ready to shove off at the first gray hint of daylight. We slept in shifts, or tried to. I may have dozed off a time or two, but only for minutes, if at all. A few hours before morning it rained, but the frequent lightning strikes revealed no enemy vessels.

So where were they? They couldn't be all that far away, but there was no sign of them anywhere. A boat like theirs, even with sails, would be invisible over ocean swells just a couple of miles at sea, so we had to keep a sharp lookout to avoid being taken by surprise. The surprise, however, was in the lack of an enemy, not the presence of it. Had they circled around the island and docked somewhere else to come at us in a land invasion?

Just before dawn, I sent two of our best fishermen in their fishing boats on a reconnaissance mission. One went left from Falelima, the other right. Their objective was to circumvent the island in search of an enemy landing party, then hurry back to report their findings. They were not to engage, and carried no weapons. Speed was their objective. That and a sharp eye. Boats can be concealed in inlets, river outlets, even in thick foliage along the coast, so they would have to investigate anything that looked suspicious.

I sent another fisherman straight out to sea, seeking the enemy boats farther out than we could see from shore. Until we had more information as to the whereabouts of the enemy, it seemed futile to launch our attack boats and send them wandering aimlessly at sea when the attack could come from elsewhere. We didn't want to tie up our only offensive capabilities in an area that might not even present a threat.

Still, it seemed probable that the invasion would come at us first from the sea. A land invasion would require them to beach their boats far away, then travel over rough terrain to get here without being seen. It might be possible, but certainly not easy. Marching a group of warriors long distances through an unknown jungle would require them to carry food and water as well as weapons. Besides, it isn't easy to sneak through the jungle without making a lot of noise.

In the afternoon, the fisherman who went out to sea returned with a report that he had found no sign of enemy craft no matter how far out he went. There just wasn't anything out there, according to him, which was a relief on the one hand, and worrisome on the other. It meant there was no attack coming in the near future, which was good. And it meant we had no idea where the enemy was hiding because it seemed they had to be hiding, which wasn't so good. They could be sitting way out at sea, beyond where we could see, waiting for—what?

Waiting for the assassins to return, maybe? Possibly, but the boat the assassins had come in couldn't have traveled that far from the main group. Nor would they plan on going that far back. No, there was something missing, something we didn't know but desperately needed to. There was an enemy out there who wanted us dead. Based on what the assassin said, they were likely cannibals, which might affect their attack strategy, the one thing we had no knowledge of.

In planning for a pending war, you try to understand the mind of the enemy. But how can you understand the mind of a cannibal unless you're either one yourself or have had previous encounters sufficient to give you an insight into their minds? The assassin had called The People animals, a herd to be fattened up. They were to be bred to produce offspring, presumably so that the herd was never depleted. Ranchers in my time treated their herds in the same way.

Counting the women and children, we numbered somewhere over two hundred, mostly women and children. Compared to the one hundred we'd been told were coming against us, we outnumbered them two to one, although only about forty of us were able to fight. That meant that even if they killed all of us who were capable of opposing them, there would still be a "herd" of about a hundred and sixty. Logically, then, a cannibal would want to conquer his enemies but not kill them all in a broad frontal attack. That would result in the loss of their intended food source.

For them, the best option would be to enslave us, killing only those absolutely necessary in the initial confrontation. Then they could manage the "herd" and butcher "animals" over time as need dictated, thus maintaining a constant source of meat. That seemed to be the logical course of action for a cannibal, but were these people logical? Were they casual cannibals, eating human flesh mainly for ceremonial reasons, or did they rely on it as an essential food source?

Some cannibalistic societies, according to articles I'd read by sociologists who studied them, ate their enemies to absorb their energy force, not necessarily as a food staple. For them, the quality of the meat was of little importance. It was the stature of the person being eaten that mattered. A chief or warrior was preferred, while women and children provided little energy. The braver the warrior or chief, the greater the energy absorbed. Was that the kind of cannibals we'd be facing?

In the final analysis, it didn't really matter what kind of cannibal they were, or if they were indeed cannibals at all. What mattered was that we be prepared for the attack when it came, ready to repel the invaders or defeat them outright. And that, of course, would be easier if we even had a slight clue as to where they were or when they planned to attack. It was clear the assassins hadn't traveled a very long distance in their canoe, which suggested that the main force was somewhere not too far away. But where? None of our efforts to locate them had been successful, so our carefully planned defenses were still pretty much up in the air.

We were getting together with the chiefs and foremost warriors on a regular basis, intending to update them on the situation, but lacking any updates to pass on. The two men we'd sent to circle the island wouldn't be back for another day, maybe two, so we didn't even have an insight into the potential for a land invasion. We had to be ready for an attack from the sea or land, and we didn't have enough fighters to protect both adequately. That would be spreading our forces too thin, no matter which direction the attack came from. No matter what we did, we were leaving weak spots in our defenses.

Waiting around without knowing what we were waiting for finally got to Sina, who didn't deal well with boredom. "I go to Tufutafoe," she announced suddenly, as if it had just occurred to her in that instant, which it hadn't. "I take ten warriors and see if bad mans is there. If no, we come back."

I didn't want to spare ten warriors, which would deplete our forces at Falelima, but I knew it would be a waste of time arguing with her, despite the fact that we'd just sent men that direction looking for the enemy. She had to do something whether it was right, wrong, or indifferent. I understood that. In that same frame of mind, I was tempted to launch our boats despite the fact that we had no target to go after. Both of us felt an overpowering need to do something. Anything. There was a war on, but so far it was a war without an enemy. Sitting around waiting for nobody-knew-what was getting less and less acceptable. There are times when you just have to take action, even if you're not certain it's the right thing at the moment.

Sina was at that point, and I was getting there as well.

She left with the ten she picked late in the afternoon. I hoped they'd make it to Tufutafoe before dark, since the trail could be treacherous enough in daylight. She chose seven men and three women to go with her, the ones who'd proven themselves with weapons and, more importantly, obedient to her orders. Maybe they didn't look like it, but Sina regarded them as a valiant fighting force.

They'd been gone for less than a half hour when the wind brought the unmistakable sound of Sina's war cry over the trees and jungle. If we were nervous before, that really put us on edge. What was I supposed to do now? Take reinforcements to back her up if we could get there in time? Beef up our defenses at Falelima for an anticipated attack? Doing nothing was out of the question. We had to do something. But what?

Taking another group of selected warriors to assist Sina would be a dangerous move. If she'd encountered the main force, she had probably been killed with those who went with her, so another small contingent would meet the same fate and leave Falelima's defenses that much weaker. On the other hand, maybe she'd come across a small force, forward observers, or more assassins. In that case, we had to assume her team had prevailed and there was no need for reinforcements.

Deciding to wait, to send no warriors to aid Sina and her group, was perhaps the hardest thing I ever had to do. I didn't know if she was alive or dead, injured, captured, or whatever. I wanted to lead a team to her rescue, or at least to find out what had happened, but I just couldn't risk the safety of Falelima. We were not an experienced army, hardened on the battlefield and ready for action. This was an army of fathers and mothers, farmers and cooks, and artisans in numerous disciplines, none of which included the bearing of arms.

That was probably the longest night I ever tried sleeping through, to no avail. I worried about enemy forces attacking, a concern that was even greater after hearing Sina's war cry, and I had to worry about her. Was she alive? Safe? Dead? Wounded and in need of the help I wasn't sending? Was I doing the right thing by keeping our forces in Falelima, or was I betraying my best friend? It seemed like no matter what course of action I took, the chances were good someone, if not everyone, would suffer for it.

The rain that night was like an extension of my mood, dark and dreary. I had terrible visions of Sina, lying wounded and helpless, drenched by the rain and shivering, waiting for the help that wasn't coming. How could I know if I was doing the right thing, or if I was screwing up royally and putting my best friend, and maybe all The People, at risk? Too much had been put on my shoulders, and I wasn't up to the task. I was a mother, a teacher, an archaeologist, not a military general. How could I be expected to make life or death decisions like that? And why, for that matter, were The People willing to follow me when I knew nothing more about such things than they did?

The morning sun didn't change anything. I was taking a calculated risk by keeping my fighting forces in the village, despite not knowing where the enemy was. Sina had almost certainly encountered them, but in large numbers or small? If she'd run across the complete enemy force, she and those with her were probably dead, and the enemy was coming our way by land. If it was just a small landing party, Sina and her warriors had hopefully defeated them and would report back to us sometime in the afternoon.

It was a terrible thing to think about. If Sina had been killed, she'd never return to her daughters in her own time, and they would never know what happened to her. Did they even know she'd been called through time again on a mission that was supposed to have been peaceful? We were brought here to support a resettlement of The People, but now it had turned into a violent war. A war that had to be fought, but for what? These villages? Or was it more than that? We didn't even know for certain what the enemy really wanted. Did they want us as a food source, or was it the favorable climate and established villages they were after?

For me, the situation was a bit different, but no less treacherous. My family was with me, so they knew what we were up against. Apu, Little Sina, and Masana. They'd come back with me based on the promise of the Little People that this was not a wartime mission. Were we all destined to die in a time and place that wasn't ours? The thought of them dying here was hanging heavy on my mind, and I wanted to do whatever it took to keep them safe, but what was the right action to take to see to it? So much was at stake, with so little information to help make the right decisions. Was I helping to defend them and The People, or was I putting them at greater risk with the actions I took?

The pressure was more than I wanted to bear, but there wasn't anyone to pass it off on. Not even Sina, who was normally willing to take on any warrior-type stuff. This, she said, was something new and more intricate than she was willing to deal with. She'd handle the training of her would-be warriors on weapons and fighting skills, but the strategy we'd be relying on was too much for her. Why she thought I could do any better was a mystery I couldn't explain, but there I was, stuck with all the command decisions. And floundering with no clue as to what I should be doing.

Early that afternoon, one of the warriors Sina had taken with her returned with a report. They had encountered a force of six enemy troops. All of them had been killed, but in the process, one of Sina's men had died and another was wounded badly. She was staying in Tufutafoe with her seven remaining warriors to defend against any further arrivals by the enemy. She also reported that they had found the body of the fisherman I'd sent that way, or what was left of his remains after the bastards had finished cooking and feeding on him.

So they were actually cannibals. Suspecting them as such had made me mad, but this report made me a fuming mass of hateful weapon-wielding warrior. I vowed that they'd pay, every last one of them, in the most agonizing way I could conjure up for them. That, of course, meant very little at the moment. We had to kill them anyway, which had long since been decided, so what could we do that would be worse than that?

Before we could do anything, however, we had to find them.

Just before sunset, sails were sighted over the horizon straight out from Falelima. The war had finally arrived, and Sina and her forces weren't here to help. It was up to our defenses now, and our attack boats. It meant a nighttime engagement, but that couldn't be helped at this point. We had to go after them in the dark, or wait for them to land on our shores before dawn and face a land battle for which we were woefully unprepared.

Chapter 11

Things had become a bit chaotic after word spread about Sina's confrontation with a small land force of enemy fighters. The fact that they came from the direction of Tufutafoe, and were in reality a tribe or nation of cannibals, caused quite a stir. Then came the sighting of sails over the horizon, which agitated things considerably more. But it wasn't chaotic, despite how it looked. In reality, it was much more controlled than that. Men and women were scrambling to get their weapons and report to the station they'd been assigned to in preparation for an imminent attack. I was hoping beyond hope that the initial attack would be ours, aboard the boats we'd prepared for that purpose. Within twenty minutes or so, both boats were manned and ready to shove off.

Fires were being lit in the trenches made for that purpose, which had been covered with dry thatching to keep the flames from being seen by the enemy until they were on them. That was especially crucial in a nighttime engagement, which was upon us. By the time we got out to where the sails had been seen, it would be completely dark. Hopefully, that would work in our favor.

The sails had been spotted by a lookout on top of a cliff about a hundred feet above sea level, but from where we were at water level, we could see nothing. We were told the enemy boats were straight out from the village, though, so that's where we were going. Sina had taken two of my "marines," so I had to quickly scout out replacements. The woman who'd returned with Sina's report volunteered despite her long jaunt through the jungle, and another young man, eager for a fight the way young people can be, joined as well. The woman I took on my boat, the man was assigned to the other one. Then we launched.

Everyone seemed anxious to get underway. Those on the paddles started off furiously, but I told them to slow down so they wouldn't be worn out before we made contact. Besides, we wanted it to be completely dark so they wouldn't see us coming. The archers were itching to light their arrows but heeded my advice to wait until we were within range. The marines were milling about, fondling their spears and clubs as if making love to them. They were all ready for a fight, as if they thought themselves impervious to danger.

I was crouched at the front of the first boat as we crossed the swells of the reef and made our way to the deep ocean. The other boat stayed to our left, about fifty feet away. Dusk settled over us, and it became clear we might have a hard time keeping track of each other in the dark, something we hadn't prepared for. Too late to worry about that now. We were on our way, and the crew smelled blood. I just hoped it wouldn't be ours they'd be smelling.

Just before full darkness fell, we spotted the masts of the enemy "fleet." They were floating close together with sails down, making it difficult to get an exact count as to how many boats there were, but I was pretty sure I counted six. They were about a mile away at that point, but close enough that we had to move ahead silently. All talking ceased, the paddle master quit calling out a rhythm, and care was taken to avoid excessive splashing with the paddles. The marines held their weapons tightly, not wanting to drop them and alert the enemy to our presence. Archers waited with knocked arrows, tipped by dried coconut husks dipped in whale oil, anxious to light them and burn an enemy boat.

The enemy was careless. There were torches and fires aboard their boats, guiding us right to them. When we'd closed to within a hundred yards, the other boat, little more than a ghostly shadow on the surface of the water, veered farther to the left to come in from the other side in a sort of pincer move as planned. At seventy-five yards, I heard laughter from one of the boats we were closing on, and at fifty yards, voices. They didn't know we were on them, and my heart leaped with joy.

At thirty feet, I whispered for the archers to light their arrows and open fire. They did with surprising accuracy, and shouts erupted on the first enemy boat as flames took hold of the reeds that formed its perimeter. At first, it was shouts of surprise, then shock, then outright panic. A man trying to extinguish the flames fell with a burning arrow in his side while another one tried to see through the darkness to locate the attack. There were no others on the boat. Just the two.

Turning to attack the next boat, we discovered the same lack of warriors aboard it. Two men, both of which were riddled with arrows while flames licked away at their craft. Farther away, our other boat had set another enemy craft ablaze, revealing once again no host of warriors. We both hit the fourth boat at the same time with the same results, and then we could find no more enemy craft. I was sure I'd seen at least six boats before darkness fell, but now we could find only four.

We'd been warned that a force of a hundred or so warriors were coming to kill us. Two of them had been the assassins, six more taken care of by Sina and her crew, and now eight aboard the boats. That totaled only sixteen, which left over eighty unaccounted for. And I'd seen six boats before it got too dark to see, but had found only four. It didn't take but a second to realize we'd been outsmarted, that a force of over eighty killers were headed somewhere in at least two boats. But where?

If they'd come in toward Falelima we likely would have heard or seen them on our way out. Was it possible we could have missed them? Yes, of course. But not likely. Which meant they'd probably gone somewhere else. I whirled on the woman who'd gone with Sina.

"Tell me about the men you fought, Tasipo."

"They try to fight, we kill them."

"What were they like?"

"Very dark, like the skin of midnight."

"Weapons?"

"Three men had clubs and knives; three men only had knives."

"So they weren't prepared for a battle, were they?"

"Sina said they were stupid men. Go to war but forget weapons."

"How old were they?"

"Young. Boys, they were, but fought plenty hard. Killed Usipe and broke the arm of Tesepe."

"Did they attack you, or did you attack them?"

"We heard talk ahead and sneaked up on them very quietly. Took them by surprise.".

"That's what I thought," I sighed, then turned to Imenoa, the man on the rudder of the boat. "We go to Tufutafoe now. Go very fast. Enemy men go there. Sina is there with only seven warriors, so we must get there in time to help." I turned to those manning the paddles and explained it to them as well. There was so much noise, between the cries of men on fire and the crackling of flames that leaped ten feet into the air, that those in the other boat couldn't hear me, so I ordered my crew to take us over to them.

Fifteen minutes later, both boats were headed toward the coastline, which we'd have to follow to find Tufutafoe in the dark. Sina wouldn't have any fires or torches lit, so we were going to have to hug the reef in the hopes of seeing the village. I prayed that we wouldn't find it covered with the blood of our own. If we could somehow beat the enemy and help shore up the meager defenses there before they arrived, we might yet save the day. But they had a substantial head start on us, so it didn't look good.

I was sick with worry. Sina and seven unseasoned, so-called warriors against eighty killers, cannibal killers, no less. They wouldn't stand a chance unless we got there in time. And that didn't sound all that promising. Our boats were slicing through the water at a speed I knew the paddlers wouldn't be able to keep up for long, so I informed the marines that they'd have to relieve them when they got too tired to maintain the pace. And then the archers. We had to maintain a breakneck pace if we were going to save Sina. And those with her, of course, but my mind couldn't escape the thought of Sina fighting alone against impossible odds, and me not there to help. And win, or die, at her side.

We plowed on through the night, giving everyone rest breaks as needed from the paddling, keeping just outside the reef using the whitecaps and the crashing sound of the waves as our guide. Despite being so close to the shoreline, we probably would have missed Tufutafoe completely if it hadn't been for the three boats that seemed to be moored midway between the reef and shore, boats with a lot of activity on their decks, lit up by a multitude of torches.

The archers were on the paddles at the time, but those assigned to paddle quickly took their place so the archers would be free to do what they'd been trained for. The marines took up their weapons, and we all crouched beneath the shields we'd erected.

"Charge straight at them," I ordered, and our two assault boats broke across the coral reef in unison. Fortunately, it was high tide, so there was plenty of clearance over the coral, and our speed wasn't hampered. We rolled and pitched with the waves, but nobody noticed. We were hunters, and our game was in sight. The war the young men wanted was here.

The three enemy boats were literally packed with men, fighting men, who seemed to be paying attention to the shore and the malae beyond. Only a handful of them noticed us coming at first, but by the time we were in archery range, we were the center of attention.

Burning arrows lit up the night sky, some falling short or wide of the mark, but some striking home. Two men fell into the ocean with burning arrows somewhere in their bodies; I didn't have time to notice where. And the first boat we reached was starting to burn.

We came in too fast. After setting that first boat on fire, we couldn't turn away from it toward the next one. That second boat was in range, though, and the archers sent a volley of fire at it until we rammed into the first one. The impact was more than I would have expected from a collision with a reed boat, but it had a solid inner frame to which the reeds were secured. Hitting it was like running aground at full speed.

There was a loud splintering sound as wooden beams shattered, and I was relieved to see it was theirs, not ours. My men worked feverishly to pull us back off the enemy boat as black warriors fell aboard ours from the force of the collision. Others actually jumped aboard to escape the flames as much as to engage us. Most of those who fell, and some of those who jumped, were impaled by the javelins we'd attached to the front of our boat, their bodies suspended on them, dangling motionless over the water. That gave a few of the others the chance to cross over those bodies and board us.

Since I was at the front, I was the first to engage the enemy. The first one who came at me had only a knife for a weapon. Apparently, he'd been forced to jump onto our craft before he could grab his war club, which was fine with me. My spear was longer and far more deadly than his knife, a lesson he learned quickly as his innards started hanging from a severed belly. He died almost before I could cross weapons with another one.

The next one had a bow, but he was being battered about by his own men who were scampering away from our boat. The tip of my spear opened his throat while he struggled to knock an arrow, and he fell sideways into the water. That's when it dawned on me that the waves were rather high, and we were bouncing on the surface more than usual for such a large and heavy boat. There was fighting and shouting in every direction. As planned, my warriors grouped themselves in the middle of our boat, forcing the enemy to balance themselves on the outer edge. Many of them fell overboard without ever managing to get into the fight.

A huge man with a war axe came at me. I parried his first swing, although his strength and the weight of his club sent shock waves down my arms, and I hoped it hadn't damaged my spear. As he pulled back for another swing with his heavy weapon, I twirled away and to his side, bringing my spear in a full arc that opened up the sinews in his back. It did considerable damage, but I positioned myself to fend off another attack. It wasn't necessary. He dropped to the deck, limp and twitching. I meant to finish him off with a blow to the back of his neck, but didn't have time before another attacker was on me. A stab to his throat made short work of that one.

For a moment, no one came after me and I had a chance to survey the carnage. The second and third enemy boats were both on fire like the first one. Warriors were jumping overboard to escape the flames, struggling toward the shore where Sina and her few defenders stood against them, vastly outnumbered. As I watched, though, the enemy warriors who made it to shore didn't engage Sina and her fighters but ran into the jungle beyond in fear, disappearing into the night. The reason became clear as more and more enemies dragged themselves onshore. None of them carried weapons.

The heavy stone heads of their clubs and axes were effective in battle but virtually impossible to swim with because of the weight. Those who went overboard, then, were virtually defenseless when they reached land. Although I'd never thought of it before, I'd be in pretty much the same situation if I fell into deep water. My spear, made of iron wood, was too heavy to float, so in order for me to swim, I'd probably have to let it go.

I didn't plan on going for a swim at that moment, though. I went after the next black warrior I could find, cutting him down from behind before he knew I was there. Then there were no more on the boat except for those who were dead or dying. It was a gruesome scene. Seven or eight men with black skins lay about, some of them groaning in agony and clutching wounds, others not moving at all. Between them were war clubs and axes, strewn across the deck as those who wielded them fell onto the boat or overboard.

One of my archers was among the dead, his skull crushed. A marine's jaw was smashed back into his head, killing him. Another had a broken hand. Those who'd manned the paddles weren't spared, either. One was dead from an injury I couldn't make out, another working to stop the bleeding from a leg wound. I wasn't sure if he'd survive or not, but there were others helping him, so I ordered those healthy enough to paddle to get us to shore.

It's odd that I hadn't heard it before. Or maybe I had, but it just didn't register in my mind amid the turmoil of the fight. Sina's war cry, rending the night air, but this time more in frustration than anger. Every time she tried to engage an enemy, he ran off into the jungle, leaving her spear unstained by the color of blood. To a warrior like her, that was an insult only a coward could give. She wanted a fight in a big way and wasn't getting it. Those with her, I think, were grateful to avoid a fight with so many of the enemy, but Sina was enraged.

When our boat slid onto the sands of the beach, we disembarked at the ready, then watched as dozens of dark-skinned enemies dragged themselves from the ocean and scurried off into the jungle, weaponless. The battle was ours, a total rout of the enemy. At least for the time being. The enemy was still on our island, near our villages, and we had destroyed the boats in which they might have gone back to their own lands. They were stuck, landlocked right where we didn't want them.

When Sina saw me wade through the twenty feet of water from the boat to shore, she came over quickly. There was no hug, no thank you or congratulations, just a huffy snort. "It is not a good war when bad men will not fight," she spat. "Can you kill a man who runs away in fear? No, I say. This is not a good war, Salini."

"No," I nodded, thinking that no war was ever good but knowing better than to say that to her, "but it is in my mind that it is not over. The enemy has lost its weapons but is now on our island and has no boat with which to leave. It will not be easy to hunt them down and kill them in the jungle, and they will make new weapons and come to kill us."

"And eat us," she agreed.

That was a chilling thought. Would they sneak around the jungle and attack anyone they found alone? In order to survive, we had to have people working in the plantations or hunting bats, and they would be easy prey. Would they take a child to feed themselves? There was still too much we didn't know about them, but we did know they'd cooked and eaten a captured fisherman. What else would they do? To what lengths would they go to kill and eat another human being?

We were going to have to hunt them down. Along the coasts, in the jungle, on the mountains, or anywhere else they might hide, we had to find them and eliminate them once and for all.

Chapter 12

Everybody on our boat crews, including those who manned the paddles, the archers, and the marines, all came together with Sina's people to create a united front. They split up into groups to take turns sleeping, since everyone needed and deserved rest. We had been up since before sunrise the day before, and the sun was just rising over the hills again. Long after the fight ended, an occasional black man would sneak from the ocean and dash off into the jungle, but Sina was tired of chasing them and just watched them go. She wasn't going to catch them anyway, and they weren't going to fight, so there was no sense in it.

The People made sure a sizable force was always on guard. There was no way of knowing how many of the enemy warriors had made it to shore and into the forest, but it was undoubtedly a formidable number. Prior to coming to Tufutafoe, I had estimated that sixteen had been killed out of the hundred or so we had been told were coming. That left about eighty, probably more, and it was hard to tell exactly how many had died as a result of our attack on their boats.

After asking those who had fought on my boat, I gathered that perhaps eight or nine had been killed outright. A few more went overboard with wounds that may have proven fatal, but we couldn't be sure. The other boat hadn't encountered any up-close fighting but claimed that arrows had taken at least a few. To be conservative, I estimated the dead enemies off the coast of Tufutafoe to be ten, possibly a dozen. Assuming none had drowned trying to reach shore when their boats were burned, that still left somewhere between sixty-five and seventy to be dealt with.

I started to tell Sina what we were still up against, but she was fast asleep on the sand. I lay down next to her, the number running through my head along with an urge to get Little Sina and the women and children back into the village. I didn't want any cannibals to discover all our helpless villagers away from the protection of warriors. That thought was my last until I woke up that afternoon.

Sina was already awake and off doing something. The people whose turn it was to sleep after standing guard for the other half were scattered all over the place. Some were sleeping on the sand, others managed to find a spot in the fales we'd built. Those who slept first were now standing guard, and with so many enemies somewhere in the nearby jungle, guard duty was no laughing matter. No matter how many of them we'd killed, there were still more of them than there were of us. Sina and her seven, plus the crews from my two boats brought our numbers to a little under thirty. Roughly half of what the enemy had.

"You are awake!" Sina's familiar voice came from a cluster of people to my right. She waded through the deep sand and looked me over carefully. "You look well, Salini. It was a good battle for you?"

"My people did well," I nodded.

"Your people say you do well. Kill four, maybe more. Say they are afraid, but you are not afraid. It is as I say, you are the greatest warrior ever."

"My people were afraid, yes. So was I. But we all did what needed to be done. And now there is something else that needs to be done."

"Yes, we search the forest for bad men."

"That will have to be done, yes. But first we must get Little Sina and the women and children who are hiding in the forest back to the village where they can be protected. This we must do before the enemy finds them."

“Aiyeeee,” she said, almost choking on the urgency she suddenly felt. “I did not think of this thing. We must go now to Falelima.”

“The boats will be faster than the trail,” I said. “We will have them ready to depart very soon. Those who are sleeping may continue to do so on board.”

It took a while to gather my boat crews, who were admittedly happy to be leaving the place where so many enemy fighters could be lurking just behind the edge of the forest. The first thing they had to do in preparing to leave was to dump the enemy fighters' bodies, those who had died on our boat, into the ocean. There was no attempt to clean up the blood, gore, and intestines that littered the surface. It was enough to get rid of the bodies to make room for our trip back to Falelima. An hour later, we were at sea again, with half of our crew getting some well-deserved sleep.

On the way back across the reef, we passed what was left of the enemy boats. They had burned down to within a foot of the water’s surface and were slowly drifting southward, smoke trailing behind them. We saw no signs of dead men but did see a few more sharks than normal, which might have aided in the disposing of the bodies. Those we threw overboard as we were leaving would likely meet the same fate, which was alright with us.

It was nearly dark again when we arrived at Falelima. The warriors were on alert, not taking any chances while we were out. They’d almost given us up for dead since we’d been gone a full day for a sea fight that should have lasted a fourth that long, and were thrilled to see us return. The first thing I did was send a half dozen warriors to the hiding place of the women and children, with orders to bring them back. They came into the village well after dark, safe and glad to be back.

Little Sina had anointed herself as the guardian over the helpless women and children and carried a spear that someone had given her.

The three warriors I had assigned as guards, she said, were too scared to be dependable. If trouble were to break out, she was afraid they would run rather than fight, and she wanted someone with a weapon that could be depended on. That was the last thing I wanted her to do. Stand alone against an armed force? Not if I could help it, but she reminded me that she, too, had made a kill, which qualified her as a warrior and guardian over those unable to defend themselves.

The next morning it was overcast. Rain was threatening from black clouds, and the wind was stronger than usual. Sina, who was supposed to be organizing a war party to go after the enemy forces at Tufutafoe or in the interior of the island, wherever they were, came to me a short time later.

"This is bad wind," she said. "Very bad. We do not hunt this day."

"How bad will it get?" I asked, almost forced to shout in order to be heard.

"Very bad."

"Are we in danger?"

"Maybe it is so," she replied with a nod.

"So where should we go to be safe?"

She shrugged her shoulders in a helpless gesture. "Where is safe from wind? There is no such place."

"Wouldn't we be safer inland?"

"Trees fall there. We stay here."

A half hour later, the hurricane hit with full force. The village now consisted of nearly two dozen fales, but two of them were reduced to frameworks within minutes. Another half dozen were destroyed within the first hour. We put the children and pregnant women in the remaining shelters while the rest of us huddled in the sand just outside so we could evacuate them if the fale started to give. It was hard to stand against the wind, let alone walk or run, which we needed to do frequently. The rain was driven as if from a pressure hose straight into our faces, and we were soaked to the skin. Being wet from rain was nothing new, but this time the temperature dropped significantly, and we were shivering in the cold we weren't accustomed to. You don't usually get cold rain in the tropics.

Trees, growing in lava rock with shallow root systems, were falling everywhere. A coconut tree collapsed on a fale in which children were huddled, but fortunately, there had been enough warning before it actually fell to get the kids out. Petu wasn't so lucky. A tree fell on her as she was rushing to rescue her pigs. She died instantly, her head crushed. The sound of trees falling along with the collapse of structures like cookhouses and outhouses came from near and far. Visibility was limited, so most of what we heard falling couldn't be seen, but what we could see was bad enough. The crying and screaming of children fought the sound of the wind to be heard.

There was no way to escape the rain. The howling wind blew it straight through shelters, hitting with a force that felt as if we were being pelted by small rocks. Any mats or clothing that wasn't weighted down blew off into the jungle, and a lot that was. Even the fales that withstood the storm lost a lot of thatching. It was obvious Falelima would have to be rebuilt.

Late in the afternoon, the winds started to let up, and finally grew quiet. There was no way of knowing, however, if it was actually over or if we had just arrived at the center of the storm. If so, it was going to start up again when the other side reached us. Sure enough, it did, and we huddled through more hours of cold and devastating winds. The darkness of night made it seem even worse. We couldn't see what was happening, relying only on what we could hear, and it sounded bad.

Just before dawn, the storm broke, and when daylight arrived we surveyed the damage. It was pretty much complete. The village was destroyed, only four homes out of nearly three dozen were left standing, and they had suffered damage. Three people had been killed by falling poles or trees, another nine injured. Livestock, pigs, and chickens, had been driven from their broken-down enclosures and ran off into the jungle. Our crops of taro, ta'amu, bananas, and other fruits were pretty much wiped out. Coconut trees, those that hadn't fallen, had been stripped of their nuts.

It was a bleak and humble group of people who walked through what had once been our homes. This was our village, built with pride and a lot of hard work. Just as disheartening was the destruction of our crops, the bulk of our food source. Now what? Not only did we have an army of enemies somewhere on the island just waiting for the chance to kill us, but now we also had what shaped up to be a famine to deal with. Our spirits were low. It felt like we'd been defeated, not by the enemy, but by Mother Nature. And she wasn't through with us yet.

The cry went up late in the morning, and panic set in. A huge wave had been sighted on the horizon, coming fast. A tidal wave. I knew from articles I'd read that such waves travel at speeds ranging from four to six hundred miles per hour. What started as a thin dark line on the horizon grew into an ominous roiling wave, giving us little time to run inland far enough to be safe before it hit.

The jungle behind Falelima rose toward a mountain beyond, but not very high or steep. We made it to a spot maybe fifty feet above sea level before it crashed down on what was left of the village. Some of the water rushed over us, but only about knee-deep. It was powerful, though, and we had to hang onto the children to keep them from being washed away. Debris from broken fales, cookhouses, outhouses, and livestock pens washed about us, and then it started to slip back into the ocean.

The backwash did more damage than the oncoming wave. Trees, poles, and other debris, being carried toward the ocean in a frighteningly powerful flood, ripped down anything left standing. We could take solace in the fact that no one had been carried away, but the village was a total loss. Even the sand on the malae was gone, scoured down to bare dirt and rock. We had no homes, no crops, no animals, and little hope.

It was just the habit of always hanging onto our spears that allowed Sina and me to keep our weapons, but most of those belonging to the villagers were gone. So there we were, basically defenseless, facing starvation, homelessness, and worried that a war could fall on us at any time. We realized that the cannibals had probably suffered in the storm as well, but had no idea how badly. Most of their weapons had been lost during the sea engagement, so they were down to knives like us, but with knives, you can make other weapons that are far more effective.

We had that ability also. We could make weapons with the knives we had, but we also had a village to rebuild. Homes for the children and elderly, women with child. We had to build cookhouses, outhouses, try desperately to salvage crops, and find ways to feed ourselves. That was the advantage the enemy had. To them, we were their food source, which meant they might strike sooner rather than later.

The power of nature had left us in awe, destitute, bewildered. You can't fight that kind of enemy, can't even imagine the force it can throw at you. All you can do is try to rebuild when it finally recedes, but for a long time, we were in shock at what had befallen us. More than anything at that moment, we needed a ray of hope. Something, or someone, who promised a healing of the wounds we'd suffered.

Masana turned out to be that ray of hope. When we needed a leader who could turn desolation into positive action, she stepped up to the plate and gave inspiration to a people who felt defeated.

"Who are we?" she shouted over the sounds of sobs and crying children. "Are we to be broken by winds and waves, or are we The People, who have built and battled and won? Yes, we are The People. We are mighty, not meek. We have built, and we can build again. We have fought, and we will fight again. We will not go down in defeat, but will become even stronger in the shadow of this storm, for we are The People."

Her enthusiasm caught on at first with only a few, but then Little Sina added her two cents worth.

"I am only born as half of The People," she called out, "but I am with you, and will build with you, and fight with you, and help to make us strong again. It is pride I feel, being among you, The People. My people! Let us show the world that we are stronger than the storm, stronger than the enemy, stronger even than the land which is rightfully ours. It is not the land of cannibals, but of The People, and I give myself to you for whatever it is I can do."

"Yes," Akime shouted, waving his hands in the air in excitement. "Yes, we are The People, and the queen of all Mutasi is with us. We can do all that needs to be done, and more, for she guides us with grace and honor and courage. Can we be any less than our queen? I say no."

I'd suspected that Akime had a crush on Masana, and his words seemed to verify it. Still, his call for courage and the strength to follow the guidance of Masana was good, and it did have an effect on many of those who stood about, still cold and wet, still afraid, but ready to tackle the challenge given us. Then Sina had her say.

"There is much to do, but we are many and we are strong. We will begin now, and quit not until Falelima is as it was. No enemy will stop us, for we will destroy any enemy who tries. Salini, the great warrior who killed Big Tooth, and evil gods, and the Aunu'u, is with us, so we will be victorious. This is my word, and it is a good word, and a true word."

Well, that was all it took. The optimism of a few overcame the pessimism of the many, and suddenly The People were on their way to recovery. The storm hadn't defeated us, the wave hadn't killed us, and the enemy didn't scare us. We were a unit, a tightly knit group ready and willing to do the work that would make everything normal again.

I was still worried. Our crops were damaged, some ruined. We could save some, but it would be a while before anything would be ready to harvest. Our animals for meat had gone into the jungle. Maybe we could catch some of them, maybe not. In the meantime, we'd have to rely on the ocean and whatever wild foods we could find. Without a doubt, we were going to have to divert some of our fighting forces. Getting food was every bit as essential to our survival, if not more so, as facing off with the enemy.

If we could manage it, the war with the cannibals was going to have to wait.

Chapter 13

It was Atamiso, chief of the family Kusani, who came up with the idea. On the surface, it seemed idiotic to me, but several other families argued that it had worked for them in times of trouble, so it was put into effect.

The first thing it required was a huge hole to be dug in the ground, roughly ten feet square. They wanted it to be eight feet deep, but the lava rock was too difficult to cut through, so they finally had to settle for four feet. Why they built it so far from the village, over two hundred yards away, was a mystery to me. It was supposed to be our food source eventually, so why put it so far from the village? They seemed to know what they were doing, though, and were excited about it, so I kept my mouth shut.

They gathered up all the food they could find, and as soon as we had some cooking pits cleared or dug, started cooking enough for the next meal. The rest, anything more than we would eat in that meal, they took to the pit they'd dug and dumped it in as if it were a garbage dump. All of it was thrown in together. Coconut meat removed from the shell, bananas, taro, ta'amu, papaya, some wild greens, and anything else that was edible but wouldn't last was just tossed into what they called a masi pit. When it was full, they covered it with banana and taro leaves, then left it to rot. The whole thing seemed bizarre to me, and I wasn't looking forward to the need to eat out of it.

While they were working on that, Sina was helping her warriors make new spears and clubs to replace what was lost, then started training them again. This time they were focused on a land assault rather than one from the ocean, and I spent a fair amount of time training with them.

Masana had a team of men and women working to rebuild shelters and cookhouses, with Little Sina joining them. Apu wanted to go deep into the forest in search of avocados and other wild fruits that grow inland, but Sina said he'd have to wait until she had a team of warriors she could trust to escort him. There were, after all, cannibals somewhere out there, so no one left the village without a security escort, which was still being trained. He did, however, get a half dozen fruit bats from the trees nearby, which helped.

A team of eight fishermen went out to bring in fresh fish for our meal and came back with a sizable load, some lobsters, and octopus, so we didn't go hungry. Not yet, anyway. I knew we could depend on food from the ocean so long as the enemy didn't cut off that avenue, but it would be a pretty meager diet if all we had was fish, and our produce wasn't going to last long. Exactly how the masi pit was supposed to help was beyond me, and nobody bothered to explain.

For the first time, I wasn't in charge of anything. Between Sina, Masana, Apu, and Little Sina, things were moving along quite well, so I made a decision about what I could do to help. It required Sina's approval, though, so I sought her out.

"It is not a good idea," she said emphatically, then gave me that grin of hers that I always loved, "unless I go too."

"It was my hope that you would say that," I agreed.

"Tonight we go. Yes?"

"Yes," I nodded.

"But we be very careful. Very quiet. Make no sound, say no words."

"Of course."

Little Sina and Apu thought it was too dangerous and begged me not to go. I was grateful for their concerns, but it was something that

I thought needed to be done. Sina felt the same way, and we'd waited too long already to do it.

"We have to know where they are and what their plans are," I explained. "How can we protect ourselves if we wait for them to take us by surprise? We need to know what we're up against, how many of them there are, if they've made new weapons, and what we should expect next. That's so basic in terms of defense that we can't put it off any longer. They've had several days to prepare for an assault against us, so we've got to know more than we do now."

"I go too," Apu demanded.

"No. We'll be taking a small fishing outrigger that will only hold two, and we don't intend to engage them. All we're going to do is find them under cover of darkness, and see what they're up to. They won't even know we're there."

"What if you don't find them?" Masana asked. "Or worse yet, what if they discover you?"

"You'll just have to trust us. We were able to remain unseen when we were getting ready to fight the Aunu'u, weren't we? We've done this before, as you know, and I promise we'll be careful. There's no one I trust more than Sina for something like this."

We paddled out beyond the reef as soon as it was completely dark, then followed it as we'd done in the attack boat. This time there was more moonlight, so we were able to see the inlet that led to Tufutafoe without any trouble. Finding it, however, did reveal trouble. The cannibals had taken over the village, but they'd suffered some losses of structures in the storm just as we had. The difference was that they were making no efforts to rebuild. We silently paddled on beyond the village before crossing the reef and bringing our canoe to the shore at a safe distance from the village.

With our spears at the ready, we crept to a spot where we could observe the village without being seen, and neither of us liked what we saw. It was impossible to get an accurate count of them, but there were many. Some were sleeping on what sand was left after the tidal wave swept the coastline clean, while others were milling about or talking in groups. There was some laughter, but most of the voices seemed somber. We had to assume that some of them had been injured or killed in the hurricane or the tidal wave that followed, and there was evidence to suggest exactly that.

There was a man, or what was left of him, hanging from the frame of what had been a fale. Despite the darkness, it appeared to me that he'd been roasted and carved, with the remnants hung after the meal. So, they ate even their own people. Maybe he'd been killed in the storm or wave, but did that justify eating him? Or maybe he wasn't a victim of the storm, just a random pick for dinner. It was hard to believe what we were seeing, but it certainly confirmed the claims of cannibalism. This was an enemy we had to get rid of. There was no other way.

Once we were back beyond the reef and far enough away that our voices wouldn't be heard, Sina suggested a nighttime raid against them. She wanted to bring all her warriors and hit them at night, when they wouldn't be expecting it. I stopped paddling and turned to her.

"How will you get so many there without making noise that will alert them? Our large boats are not quiet. Besides, it seems that they are many and we are not. How many warriors can you bring against them?"

"Twenty," she exclaimed with pride.

"But they have twice that many. Maybe three times. It would be three against one, and they've probably had time to make weapons."

"They make stupid weapons; we make smart weapons."

"Stupid weapons can still kill, Sina. I know we need to do something, but let's think up a better plan than an open attack. Especially one in the dark."

She fell silent after that, and I was pretty sure I'd pissed her off, but a nighttime attack against those odds didn't seem very likely to have a good outcome. It was inevitable when we finally went to war that we'd suffer casualties, but I didn't want to sacrifice any more than necessary in a foolish, over-confident assault. If that made her mad at me, it was a price I was willing to pay.

But I kept paying it for days. Sina had never been so mad that she wouldn't talk to me, or do anything with me, or even look at me, but for several days she avoided me as if I had some terrible contagious disease. She even slept in the fale of Masoto, staying up late at night to talk with several others who seemed to like doing their visiting in the dark. Which wasn't unusual. That was the coolest and most comfortable time of day, so it made sense.

It should have dawned on me, though. Sina was training her people hard, putting them through an exhausting regimen, and then they sat up way late at night talking. I should have realized something was up, at least that something wasn't right, but I missed it. I didn't put the pieces together, didn't even give it a thought until it was too late. In my defense, Apu, Masana, and Little Sina missed it too.

"You not go too, Salini of Siata?" Misiapo asked as I stepped from my fale early in the morning.

"Go where, Misiapo?"

"To kill bad mans. All warrior go."

"When will they go?"

"They go when it dark last night. All warrior go."

I looked around, and the village did seem empty. There were women doing their normal morning chores, though not as many as usual, and not a man in sight.

"They go last night?"

She grinned, nodding. "Go to kill bad mans."

"How did they go, Misiapo? Did they take a boat?"

"Yes, two boat."

Panic flooded over me like a wet blanket. Suddenly it was hard to breathe, and my stomach was churning. Sina had taken her warriors to do exactly what I didn't want them to do, and I'd been too blind to see it coming. They were probably in a fight for their lives at that very moment, and I wasn't there. Sina had always made a point of how good we were in battle when we fought together, and then she'd run off to war without me. If she'd been mad at me before, it was nothing compared to how mad I was at her right then.

I caught up with a fisherman just launching his canoe for the day's catch and literally took his boat at spear point. Since I'm not as strong at paddling as Sina, or most of the others for that matter, it took three tries before I could cross the reef to calmer water, then headed for Tufutafoe with all my might. Two hours later, I pulled into a rocky shoreline about a quarter-mile before Tufutafoe, pulled the boat as far onto the land as I could, and went the rest of the way on foot, my spear at the ready.

The sounds of screaming and yelling wafted across the Trade Winds long before I came in sight of the village. I couldn't make out the words or recognize the voices, but it didn't sound like a battle. There was no clash of weapons, no war cries, no shouts of victory or defeat. Just screaming and yelling, and lots of it.

I had to crawl around some big rocks that jutted out to the water's edge, then through some jungle growth beyond before I got a glimpse of the village itself. What I saw was incredibly bad. There were bodies, probably well over a dozen, scattered all over the malae, both black and brown. At first glance, it appeared that there were more blacks down than brown, but I couldn't be sure. If true, that should have been good news, but it was what was happening in the middle of the malae that was bad.

I didn't know how many warriors Sina had taken with her on the ill-fated attack, but more of them were alive than lying dead. The problem was that they were all kneeling unarmed in a cluster, surrounded by at least thirty black men with a variety of weapons. Some carried crude clubs, others the spears or javelins Sina's warriors had taken with them into battle. One of Sina's warriors said something to his captors I couldn't hear, then started to rise. A tall, bald-headed black man swung a vicious-looking club in a wide arc that crushed his skull. I didn't get a chance to recognize him before he fell on what was left of his face. These captors were serious, and not concerned with the rights of prisoners of war. There were angry murmurs from the other captives, but Sina quickly silenced them with a single word.

If these guys really were cannibals, and everything suggested they were, they had as many meals as they could want for a long time right there on the malae. Since there was no refrigeration or ways to preserve meat, the logical thing for them to do was keep the prisoners alive until they were needed for food. But I didn't trust them to be logical, or even care about such things. I had to do something, and do it fast, before more of Sina's people, or Sina herself, got killed.

The boats Sina had brought were pulled up against the shore, and on the far one, I saw a bow and a stack of arrows. That gave me an idea, but getting to the boat meant going out in plain sight. Unless I swam to it, that is. I backtracked through the jungle brush and around the rock, hid my spear where it wouldn't be seen, then started swimming out to deeper water. Once it was deep enough, I turned back toward the boats, swimming silently but as quickly as I dared. As the malae came into view, I dived, swimming completely underwater until I was behind the nearest boat. There, I surfaced to take a breath, pulled myself along it where I couldn't be seen until I was exposed again between the boats, then went under again.

The bow had been dropped directly in the middle of the boat, and I couldn't reach that far from the water, and to climb on board would put me in plain view of the enemy. The arrows were closer and easy to grab, but without the bow, they weren't much good. I gathered them in my fist and pulled them to where I was bobbing up and down with the waves so I could get to them easily, then started to slide carefully onto the boat.

It probably took all of a minute to slither far enough to grab the bow and slide back into the water again, but it felt like hours. There was no question that if they'd seen me, I would have died, but luck was with me. They were more interested in their captives than the boat they came on. Swimming back beyond the rocks to safety with the bow over my shoulder and arrows in my hand made the trip back more difficult, but I managed it. Then I had to face the next problem. What to do next?

I had a bow, four arrows, and a spear. And well over thirty enemy warriors, so the numbers didn't evoke a lot of confidence. The bow was flimsy, not made according to my instructions, and the arrows were anything but straight. The points were whittled on the ends, the feathers poorly glued and cropped. They were designed to be used for shooting flame at a boat, not to kill a much smaller target. That was why they were left behind when the land invasion started. Flaming arrows aren't very efficient against an enemy at close quarters.

I'd hoped to find something more useful on the boat, but that's all there was. The lives of Sina and those with her were in jeopardy unless I did something soon, and all I had to help them was a bow, four mostly useless arrows, and my spear. I had no idea what I could do to help them against so many without getting myself killed or captured. The bow was made to start a fire, so that didn't help. With my spear, I might be able to take out two, maybe three, before they got me, which wouldn't help Sina. I had to think of something, and fast, but nothing came to mind.

And then I heard voices.

Two men were talking to each other, so close it was frightening. It sounded like they were just above me, in what I judged must be the middle of the jungle. I hunkered lower into my cover, worried that they were coming my way, and getting caught was the last thing I needed right then. I couldn't make out the words, but they didn't sound like they were hunting for me, which was a relief. It was more like friendly chatter with no hint of excitement, which you'd expect if they were onto me.

The voices tapered off as they moved in what sounded like the direction of the village. Encouraged, I risked taking a peek through the undergrowth I was hiding in. Yes, there were two of them, walking casually through the jungle at the foot of the hill. When we'd first come from Falelima to Tufutafoe, we'd followed that same course because it was less dense than the jungle everywhere else. But now, from what I could see, a trail had been cut through there. Enough, at least, that they could walk freely, without the need to work their way through the undergrowth.

From there, the village would be in plain sight. That's when I came up with a plan. If I could get there undetected, the bows might actually be useful. It would require some luck and stealth, but I might be able to attack the village. Maybe not kill them, but at least draw them away from Sina and her captured warriors, and perhaps even allow me to get away myself. It was a long shot, but I couldn't think of anything better. It would give Sina and her team a chance. Maybe. If everything went according to plan, that is. If I was lucky. But it was a crazy, dangerous scheme. I recognized it, registered the reality in my brain, and dismissed it again as I started working my way toward that trail.

But nothing ever goes according to plan.

Chapter 14

The humidity of the climate made crawling through the jungle quieter than the Northwest forests I'd spent most of my life in, but there were areas where I was dangerously exposed to the village. In those areas, I had to crawl on my belly, shooing away the huge insects that got in my way. Centipedes there are eight to nine inches long and have a bite that is extremely painful, even fatal in some cases. Millipedes are almost as long but harmless. Cockroaches are three inches long, and wolf spiders as big as the palm of your hand. There were hundreds of other species of insects and spiders in the jungle, which I tried to ignore, but some wouldn't be ignored.

I almost panicked when I felt something big crawling on my right thigh, but it turned out to be a millipede. I brushed it off silently and went back to crawling.

The trail, when I got to it, showed that it had been made by a lot of foot traffic. Much of the undergrowth had been cut by knives or trampled down where they'd been going back and forth so frequently. A real trail had actually been created. Spying on us at Falelima? Or maybe they just went that way in search of fruits and roots to eat. No, not likely. The jungle directly behind the village would be a much better place for that. For hunting bats and birds, too. This trail just led to the steep part of the mountain we'd had to climb to get here from Falelima.

I put that out of my mind for the moment, sizing up my own situation carefully. Yes, the village was in full view except for the malae, where almost all of them were, surrounding Sina and her crew. That worked in my favor. One small stroke of luck, at least. Hopefully, it wouldn't be my last.

Next, I had to find something flammable for the arrows. That came in the form of some soft moss that had dried up since the last rain. I wrapped three arrow tips with it, tying them on with some green vines, not expecting to have time for more than three. If that. The next problem was making a fire, at least a tiny flame to light the arrows with. I didn't have Sina's knack for starting a fire by rubbing sticks together, and she wasn't there to do it for me, so it was time to learn.

Taking the bow and arrows with me farther back into the jungle in case someone came along, I found a few pieces of wood that might work and whittled one down until it had a flat surface. Then I cut a groove down the center of it, cut the other stick until it was the right size to slide up and down the groove, and started rubbing. I knew you had to press hard and rub the stick back and forth rapidly to create enough friction to generate heat, but after several minutes of exhausting work, I realized I'd forgotten one important detail.

I piled some of the dry moss beneath the wooden plank I'd flattened, then cut a hole in the bottom of it for the spark, if I ever got one, to fall through and ignite the moss. Then I went back to work rubbing the stick over the groove. I thought my arms were going to give out on me. It was hard, painful work, at least for me. Sina made it look easy, but for me, it was anything but. Where Sina could start a fire in a minute, it took me closer to fifteen, but finally, I got a puff of smoke from the moss.

I blew on it until a small flame erupted, then quickly put the moss on a chunk of bark I found and rushed back to the trail before my fire burned itself out. Nobody was in sight as I put the small fire on the ground in front of me and lit the first arrow. The fales were lined up before me, the first one about forty feet away, the second maybe sixty. It was the second one I wanted to hit first, and since the bow wasn't very strong, I aimed high and let the arrow fly.

It went farther than expected, lodging itself in the thatch of the third fale, a good eighty feet away. The dry thatch caught quickly while I lit the second arrow and shot lower this time, catching the second fale on fire. Nobody seemed to notice yet, so I had time to light the third arrow and shoot it into the first fale. I let out my fiercest war cry, then crept back through the jungle to my original hiding spot as chaos erupted in the village.

Men with weapons sprang toward the fales, then turned up the trail to where they thought the threat came from. That left only a few guarding Sina and her crew, and it was Sina who first took action before I could get there to help. She jumped one of the guards and, with no weapon but her bare hands, wrestled his club from him and put a lethal gash in the side of his head. Two others started after her just as I dashed into the malae with my spear. For a moment, the cannibals were immobile, confused by how close I was to them as I sprinted in. They recovered quickly, though, and turned on me.

The other members of Sina's war party suddenly understood what was going on and joined the fight. Now there were better than a dozen of us against only four of them. The rest were still looking for an attack coming from farther up the trail, while a few were gazing in desperation at the burning fales. The two that turned on me kept coming while Sina took down one of the others with the club, and her other warriors took care of the other one.

Sina rushed to come to my aid, but the two black warriors were on me before she got there. I parried a powerful blow from a club, but before I could strike back the other one was swinging one of the spears taken from Sina's party at me, so I was forced to turn it aside while the first one recovered enough to bring his club at me again. I wasn't in a position to parry that one, so I dropped beneath it, which gave me a clear angle on his stomach. The tip of my spear went in a good four inches, but there wasn't time to do more because the other one was bringing his spear around again.

It hit my hip, but fortunately, it was a poorly made spear and not very sharp. It cut a gash, but not a serious one, and the man wielding it never got the chance for a follow-up because Sina's club dropped him like a rock. The one I'd stabbed was gasping for air, obviously in great pain, so I meant to finish him off, but Sina beat me to it. Those who'd been gazing at the ruined fales suddenly sprang into action, seeing the fight behind them, but Sina's warriors were on them in a heartbeat. They didn't stand a chance as her crew took up whatever weapon they could find and shielded us from further assault.

One of her warriors was struck down by a club that hit him in the back, but within seconds the cannibals were overwhelmed and fell, one by one, onto the sand of the malae. It all happened with hardly a sound, so those who were still hunting the supposed attackers farther up the trail had no idea what had happened.

One of her warriors approached Sina with her spear, which he had managed to rescue in the fight. She thanked him with heartfelt gratitude, then motioned for everyone to get to the boats. We sprinted to the nearest one, dragging the injured man with us, leaving the other boat behind.

By the time the cannibals returned from their fruitless search of the trail, we were a good thirty feet out from shore. There were a few javelins thrown at us with no effect, and a couple of club wielders tried to wade out after us, but the water was too deep. Six men were paddling furiously, and within a moment or two, we were too far out even for javelins to reach us. There was a lot of shouting from the shore, and fists raised in anger. Some of our people responded in like manner, and then we were bouncing over the reef. A few minutes later, Tufutafoe was out of sight, and we raised the sail.

"I told you it was a bad plan," I hissed at Sina when the danger was behind us.

"It would have been well for me to listen to Salini, Warrior from Siata," she replied, her head bowed in a contrite manner. "I thank you for coming to help us."

"How many did you lose?"

She did a count of those on the boat with us, then calculated in her head for a moment before answering. "Nine, I think. And Masipo will not fight again."

"Ten," I shrugged. "The enemy knows we are weaker, and they are mad. I think they will attack soon, and they have our boat to use."

"We will be ready, Salini. All is not lost."

"Ten are lost, Sina. How many of the enemy did you kill?"

"I think maybe ten, maybe more."

"Or maybe less. Either way, they are still more than we. It will not be easy to defend the village."

"Easy or not, we must do so."

"They have some of our weapons, so we must make more. Does that not make it harder?"

"We have fought many battles together, Salini. If it be only you and me, we will defeat them."

"You're being foolish, Sina. I don't know how many they are, but they are warriors who sailed for months across the great ocean to make war with us. Did they not overcome you and your warriors in battle?"

"It was a mistake to attack them as we did. It will be done differently next time."

"Next time it will be they who attack, and probably very soon. Will we be ready? Only if we work long and hard to prepare."

"You worry too much, Salini."

"And you don't worry enough, Sina. We are in danger, but you don't seem to know it."

“Then we will prepare together, you and me. And we will be ready.”

I sighed. “Yes, we will prepare together, and we will be ready.”

I didn’t share her optimism, but there was no sense arguing with her. To her, it was a simple matter in which a few warriors, most of them not well trained and without combat experience, could overtake many warriors who did. Next time, a stunt like the one that had saved her and her warriors wouldn’t work. It was going to be a fierce battle, one on one, that would decide the victor. And we were outnumbered and outclassed. It was going to take some serious preparations if we were to survive the next encounter with the cannibals.

Falelima was a somber place when we returned. There was a lot of crying and wailing over those who were lost in battle, and a sense of foreboding over our current state of danger. It wasn’t very desirable as a place to prepare for war was concerned. The atmosphere was one of defeat, The People preparing to die rather than planning for victory. If it hadn’t been for the fact that my husband and daughter were there with me, I might have felt the same way, but I had to help find a way to survive, and completely destroy the enemy. The lives of my loved ones depended on it, along with those of The People we were supposed to be helping.

I tried to call a meeting of the chiefs and the remaining warriors, but nobody wanted to talk about war until those whom we had lost were celebrated and kava shared in their honor. Sina came to my aid in that regard, sort of. She railed against them for refusing to hold a war council immediately, and stated emphatically that I had taken over the role of war chief. It was me, she told them, who would prepare our defenses against an imminent threat. That was a role I wanted no part of but found myself unable to avoid.

I had to allow the passing of kava in honor of the dead before the meeting got underway, which took a couple of hours. After the chiefs had completed that ceremonial function, complete with all the speeches that came with it, the remaining warriors were called in to join us. All in all, there were about twenty-five men and women physically capable of bearing arms, roughly half of them women. In addition, there were three chiefs young and strong enough to take part in the fight. None of them were experienced warriors except those that had accompanied Sina on the morning raid, and they had more experience in defeat than victory, so they were disillusioned about our chances.

I wasn't sure how to begin when they were all assembled. To say our chances were good would have been a lie, to claim that our warriors could prevail in a toe-to-toe battle would have been a pie-in-the-sky promise. Our only hope was to avoid that kind of confrontation with the enemy. The first order of business was the making of weapons, of course. A warrior wasn't much good without a weapon to fight with, so every possible person had to be put to work making them.

Javelins were easier to make than spears but were for the most part a one-use weapon. Once thrown, they were gone. Once in a while, they might be recovered for another throw, but by the same token, they might be taken up and used against us by the enemy. Spears remained with the warrior throughout a fight but took a lot longer to make. Since we had so little time, it was decided that the spears would have to be made without teeth. The blade and shaft would have to suffice, which limited their effectiveness. It was the teeth that most often dealt the death blow, not the blade, and I was letting our warriors down by decreeing blades only, but we just didn't have time to carve teeth into every weapon we made.

Strategic planning was the hard part, but there was a precedent that came into play. I reminded Sina of a former battle against the Aunu'u, one we'd fought thousands of years earlier in another time but at the same place. That's when she took heart. Suddenly, she was with me, helping me plan out the preparations that were needed and the placement of armed combatants. As her enthusiasm grew, the chiefs and warriors began to feel more optimistic. Once again, we had a people who were united behind a plan for survival and ultimate victory.

So much would depend on luck, and I wasn't as optimistic about that as they were, but at least we had a plan. It wasn't a good one, nor one that gave us an odds-on advantage, but it was a plan that encouraged them to work hard in preparing to put it into effect.

Chapter 15

Time was the most critical issue we faced. Because we expected a raid to come sooner rather than later, we had to take that into consideration in the making of weapons. Javelins were rather simple to make. A limb, long enough, strong enough, and narrow enough, just needed a point carved on the end. We gave up on the need to peel the bark and smooth the limb, going instead with rough, crude throwing weapons. I instructed the warriors who were training with me and Sina that they should have three or four javelins but keep one for hand-to-hand combat. It might not have the capabilities of a spear, but it could still be a good weapon for blocking blows and stabbing opponents.

The spears were a different story. To make a spear, you needed the trunk of a small tree, wide enough to carve a blade at its tip, four to five inches in width, and the staff had to be cut down enough to be easily managed by the holder. That required a lot of carving. The women who weren't able to fight could help make weapons, so I assigned only six to the task of making spears. The rest, closer to twenty or so, were put to making javelins.

One of the problems we faced was in the burying of weapons as we'd done in the fight against the Aunu'u. The tidal wave had washed away most of the sand on the malae, and digging in the rocky surface beneath it was out of the question. Other hiding places had to be found, places that would be inconspicuous yet readily obtainable. That meant hiding extra weapons in the jungle undergrowth surrounding the village. It wasn't ideal, and not as quickly accessible as if they'd been buried in the sand where the fighting would take place, but it would have to do.

Virtually everyone in the village was willing to fight, but we exempted young children, women heavy with child or nursing mothers. Some children, as young as early teens, however, joined in the training with Sina. Some in that same age group, however, with physical conditions that would hinder their ability to fight, were refused training even though they wanted to join. We didn't need a slaughter of those with bad legs or arms, or with mental faculties that would make them slow in battle. So it was that our fighting army grew to almost forty.

The youngest ones were to be held in reserve, joining the fray only if needed. They could throw javelins from the cover of the jungle but were warned against joining in close combat unless our primary forces were being cut down too much. The older teens and men were given explicit training on how to fight in conditions like we expected to encounter. They had to know as much about how to avoid the blow of a club or spear as how to strike a blow of their own. In order for a warrior to be successful, he or she needed first to stay alive, and the clubs the enemy used were lethal weapons indeed.

One of the women from the Kusani clan was doing her best to tend to Masipo, the man who suffered a club blow in our escape from captivity, but his back was broken and there was little she could do. I helped her put his back in a restraint, but he kept trying to get up, vowing he could still fight. He died late that evening, clear evidence of how dangerous the enemy clubs could be.

Guards were placed along the coast in case of an attack by sea, since we'd abandoned one of our boats. Others were placed inland in case they came at us by land. Work went on well into the night, and would have gone all night if I hadn't called a halt. If we were to face the enemy in the morning we'd need rest. Exhausted warriors wouldn't be as effective as rested ones, and those who weren't fighting needed to get to their hiding place quickly.

Morning broke with no sign of trouble, so work resumed on the making of weapons and training. Little Sina wanted to fight, but I insisted that her role as guide and protector for those who couldn't fight was more important than adding one more fighter to our ranks. Apu, however, was training along with the others, which worried me. He wasn't a warrior; he was my husband, and I vowed to do everything in my power to protect him when the time came. Assuming I could do so, of course. In battle, you're not always in a position to station yourself in the defense of any other person no matter how hard you try.

By nightfall, we had well over a hundred javelins and nine spears made. The spears were a bit rough, but functional. With the spears we already had in the village, that made a total of fourteen. We hoped we'd have more before they were needed, but those fourteen were given to the most promising of our fighting crew. Sina trained them hard, teaching them every move they might need, both in defense and offense. One was as important as the other, and Sina gave both a lot of attention.

Everyone was starting to feel more confident, so I warned them against assuming it would be an easy victory. The enemy was strong and well-trained with their weapons. They weren't going to be easy to defeat, even if we outnumbered them. I reminded them that many of our fighters were too young to have the strength and stamina of an adult or the experience that would be needed on the battlefield. We were a ragtag army, made up of fishermen, farmers, hunters, and laborers. None of them were seasoned warriors like the enemy. We had to go into this fight knowing that our lives depended on defeating the cannibals, and realizing that it was going to be an uphill fight. Taking pride in our numbers could easily get us all killed.

We had destroyed the enemy's boats, so they couldn't sail away and leave us alone. They were stuck here, and we were stuck with them. They had the boat we'd abandoned, of course, but it probably wouldn't hold them all. Even if they didn't mind being cramped, it was a long journey back to their former lands. It dawned on me that they might actually try to commandeer more of our boats. If it meant they'd leave, I'd be willing to give up a few, but they weren't mine to give. Nor did I believe for a minute the enemy would actually go away. They'd come a long way to take our lands and kill us, and I didn't think they'd change their plans now.

The next day we almost doubled the number of javelins and added six more spears to our arsenal. That made twenty spears, plus mine and Sina's, and enough javelins to put up a good fight. My concern was over a fight at close quarters. Our warriors would do their best, but would it be good enough? If we could reduce their numbers substantially at a distance with the javelins, we might have a chance. If not, it was going to be a close one.

Sina was more optimistic than anyone and started talking about another direct assault with the added strength we had. I told her no in no uncertain terms, but I was starting to wonder what needed to be done myself. We'd expected an immediate response after the humiliation they'd suffered after Sina's attack, but three days had gone by with nothing happening. Maybe they weren't planning an attack at all. How long could we wait, expecting to be attacked, if no attack was coming?

The reality of the situation was that we were on a small island, and a clan of cannibals was on the island with us. Their objective was to kill us, so there was no way we could share that island with an enemy that wanted to roast us. Literally. Either we had to go, or they did. It was that simple, and we had no intention of leaving. This was the island inhabited by The People from the very beginning, the land they would inhabit in the future when I would come here as a rookie archaeologist. Which meant this wasn't their island and they had to go. Or die. There could be no other choice.

I entertained thoughts of taking another of our boats and leaving it at Tufutafoe in the hopes that it would encourage them to leave. They needed to go to the islands far to the west. Islands that their descendants would occupy in my time. I was pretty sure they were the precursors of the Micronesians or Melanesians, or the natives of somewhere in Africa. All I knew for sure was that they weren't indigenous to the islands in what would become Polynesia. Those islands were spoken for, and these cannibals weren't invited.

It was, to my chagrin, a war over real estate, something I frankly didn't believe in. It wasn't anything new, of course, for history was full of such turf wars. What bothered me was that I was not only involved in one, but actually taking the lead role in it. Was a piece of land, or an island, worth the price? Did it justify the lives that would be lost? But this one was different in one major aspect. These were cannibals, here to kill and eat us, not just take our island. Lives were going to be lost anyway, so we had to make sure it was their lives, not ours, that were snuffed out.

Another day passed with no sign of the enemy, and I was starting to get nervous. They were going to attack. That seemed logical. They'd been the victims of an assault and humiliated by a surprisingly effective escape effort, so they had to be mad. They had to want revenge in the worst way, but how and when? What could they possibly be scheming up that would take this long to put into action?

More importantly, what should we be doing to defend ourselves against whatever it would be? Were we doing everything possible, or was there something else we needed to do? What had I overlooked that might turn out to be the key to success? These questions were weighing heavy on my mind as night fell once again with no sign of aggression.

What bothered me the most was that The People were getting complacent. They were beginning to doubt that an attack was actually going to come, that the enemy was too afraid of us after what we did to them. If they did come, they shouldn't be hard to repel. That foolhardy attitude could easily be our downfall. I knew from experience and history that the Polynesians had occupied these islands, and hundreds more like them to the east, west, north, and south, for thousands of years before my time. The darker-skinned people took the islands that became Micronesia and Melanesia much farther south, beyond the islands of the Polynesians. Some of them were cannibals, most weren't.

That made this situation all the more confusing. They weren't supposed to be here, so if history can't be changed, we had to be victorious in this conflict. But the Little People had promised us a mission devoid of conflict, which meant there weren't supposed to be any enemies here in the first place. Then again, the Little People weren't always right, or didn't see things the way we did. I had sort of a love/hate relationship with the little bug-eyed creatures. They were useful at times, a major pain in the ass at others.

At first they insisted that time couldn't be changed, that the present was someone else's past or future. Then they thought maybe it could be altered, at least in small ways. Apu and Masana were both living in my time, but originated in very ancient times. That was a change, though not on a major scale. But the presence of black-skinned cannibals on what would be known in my time as the island of Savai'i, Samoa, clearly Polynesians, was no insignificant change. What was happening here shouldn't be, and the only way we could rectify it was to kill them, or drive them back to the islands that would rightfully be theirs in the annals of time.

Driving them away didn't seem any easier than killing them, though, and that wasn't going to be easy. If we faced them in a war, it would result in a lot of lost lives on both sides. But I didn't think the threat of defeat would make them leave. I thoroughly expected that we'd have to cross weapons with them. Even if we decimated their ranks, though, I had a feeling they'd keep on fighting to the last man. Even if we provided them with the boats necessary for the journey, I thought they'd stay and fight. No matter how I looked at it, we were in for a fight to the finish.

A fight that was clearly going to be for our very survival.

The next morning the question was answered. About an hour after daylight, a logo, the hollowed log that served as a drum, sounded from the jungle above and south of the village. That was the warning of a land assault. A few minutes later, another one sounded from the shoreline, indicating an assault by sea.

They were coming at us both ways, obviously using our boat after all. Not to leave, but to attack the coast while other forces came at us along the inland trail. The time had come to win the battle and live, or lose and die.

Chapter 16

There was instant action, almost amounting to pandemonium, as everyone raced to their assigned positions. Unfortunately, the assigned positions weren't intended to defend against both fronts, inland and coastal. We had planned for an attack via the overland trail from Tufutafoe, with little concern for a water assault since we'd destroyed their boats and didn't think they'd want to use our heavier one. What we'd failed to consider was the possibility of needing to defend both a sea and land front simultaneously.

Little Sina herded the helpless women and children to their refuge deep in the jungle, while everyone else was taking up weapons and getting into position for a fight. Sina barked orders like a combat general, trying to create a defense against a two-front assault, which wasn't easy. Her warriors had to be divided, some to cover the inland assault while others were sent to the shoreline. There was a lot of confusion, but each and every one of those people did exactly what was asked of them.

Sina headed up the troops going inland while I was in charge of the coastal force. Chiefs who were normally in charge of everything related to their clans gave up that authority to me and Sina without question. I thought about putting my warriors on a boat to meet the sea attack out in the water, but abandoned the idea. It had worked before, but then the enemy was in reed boats that burned easily. Not so now. And their bows were more effective than ours, so we'd probably get cut down before we could engage them in close combat. No, we had to meet them on the shore.

"I'm with you, Mom," Masana said, surprising me when she came up from behind. "Give the order and let's kick some ass."

I was about to tell her to head up the teens who were waiting in reserve, but just then a boat laden with black warriors came into sight. I was surprised to see that there were only six of them on the boat. That was a pretty small assault team, especially since it took all of them to paddle the craft. Still, they kept coming, shouting insults and occasionally waving their paddles in the air as if they were weapons. One of them dropped his paddle on the deck and took up a bow.

"Cover!" I shouted. We were a good target, and well within his range, but his first shot went high, passing over our heads to fall harmlessly to the ground behind us. By the time he was ready for another shot, we were under cover, so he had nothing to shoot at. He fired off another shot at the shoulder of Anositi, who was partially exposed, but again his aim was bad, and the arrow missed. Anositi quickly corrected his position to avoid presenting so much of himself as a target.

When they were about fifty feet out, they began paddling furiously to force the boat as close to the shore as possible. "Now!" I cried, which signaled our archers to fire. Four arrows took to the air, followed by four more a few seconds later. One of the enemy fighters was hit in the throat and fell overboard after a minute of trying ineffectively to pull the arrow out. I'm pretty certain he bled to death in the ocean in mere moments, but that didn't slow down the remaining five.

"I thought there would be more," Masana said, standing behind me under the cover of a coconut tree. "They can't expect to beat us with just these guys."

I thought about that for a second, and a cold chill ran down my spine. "It's a trick!" I hissed. "A diversionary tactic to split us, and it worked. Half of our fighters are here, way more than we need for these guys, but the other half may not be enough for Sina."

Just as I said that, the black warriors stopped paddling toward shore, trying with everything they had to turn the boat back out to sea. They never intended to come ashore and fight us with such a small force. It was just a trick to weaken our inland forces.

"Masana, you stay here with six warriors in case they come in. Kill any you can. I'm taking the rest to help Sina." At just that instant, Sina's war cry pierced the air from somewhere deep in the jungle above the village. I quickly assigned six men to stay with Masana, and ordered the rest to come with me. They didn't hesitate, even though they knew we were going into a battle that could cost them their lives.

We sprinted through the village to the inland trail, then followed it toward the sounds of battle coming from somewhere up ahead. There was shouting, and the sounds of wounded men crying out, but which side they were on was impossible to tell. As we got closer to the fight, we could hear the occasional sounds of clashing weapons and the grunts of fighting men, but not as much as I would have expected. Then we crashed through a thick section of jungle growth and found ourselves right at the fight.

It was a mess. The jungle was so thick and the trail so thin it was impossible to spread out. Sina had selected an excellent place to meet them, which came as no surprise. The attackers were backed up along the trail in single file, the jungle so thick they couldn't maneuver. A few were trying without success to move from the trail to break through to Sina's band, but it was pretty much one at a time on both sides of the engagement. Only one or two could reach the front, and Sina's crew could only get one or two to defend against them. The rest were behind those in the lead, unable to do anything.

The attackers had expected to get into the village, where the battle would be fought in the open, and were taken by surprise on the trail. I told my archers to climb up the side of the mountain as much as possible and shoot their arrows into the back of the enemy lineup. Shooting those directly in front might endanger our own warriors, so I cautioned against that. Climbing that hill was anything but easy, but four of them managed to get just high enough to shoot over the heads of our people, and then they let their arrows fly. It was close range, so close it was hardly necessary to aim. Two of my archers shot the same man, putting him out of the fray along with two more that took arrows. They fell into the jungle, but those behind them merely climbed over their bodies and kept trying to move forward.

The archers fired another round, and three more attackers were either wounded or killed, but there was a raging pack of fighters behind them who kept trying to force themselves forward. I frantically studied the jungle, trying to find a way for my men to help, but it didn't look promising. Sina had encountered them on a very narrow stretch of impenetrable jungle, so all we could do was wait behind the others in case we were needed.

Although I couldn't see her through the mass of bodies at the front, I saw Sina's spear flash through the air time and again, which told me she was the one in direct contact with the enemy. That scared me. In a situation like this, the one in front didn't usually last very long. It was a battle of attrition, in which the one in front would go down and be replaced by the one behind him—or her.

"Javelins!" I shouted, and the men who carried them knew what to do. In seconds, crudely made spears sailed over the heads of Sina and our men. There was no way to aim, for the enemy was out of sight, covered by our people and jungle growth. But it was possible to throw over the heads of our fighters, which put them within the ranks of the enemy. I don't know how effective they were, but I heard cries of pain coming from beyond the fight. And then someone shouted a command, and as if a switch had been flipped, the fighting stopped.

There was the sound of jungle vegetation being torn aside as men scrambled away, a sound that receded like an echo as the enemy fled. Within a few moments, there was nothing but the sounds of heavy breathing from our own fighters. Sina forced her way through the jumbled warriors, coming straight to me. The blade of her spear was crimson in color.

"You have done well again, Salini," she stated sarcastically. "The enemy has been driven away, which is good for this moment, but now we must face them again another time."

"Are you well, Sina?"

"I am well, but the war is not over. They will not be as many next time, but they will come again." There was a dramatic pause before she went on. "Unless we go to them first."

"Did we lose any of our warriors?"

"One," she said with a sad nod. "But they lost many. Some by my spear, some by your archers and javelins."

That's when I pointed at the blood staining her spear. It wasn't the first time I'd seen such a sight, on her spear and mine. And many others in battles we'd fought before.

"I wanted to help, but I couldn't get in there."

"It was smart to enter the fight through the air, Salini. It broke their will to fight, so they ran away. But there are still many. How did the battle go on the shore?"

"There was no battle there. It was a trick to split our forces, and it worked. You had only half of our warriors, and this is where the fight happened."

"It is my thought that they wanted to get into the village and fight there, but that is why I picked this spot to wait for them. Only one or two at a time could fight us here."

"And you had to place yourself right in the front, didn't you?"

"I was the leader. Is that not where a leader should be?"

"And you were deprived of their blood when they landed, so you didn't want to be deprived again," I chided, knowing it would do no good. "But I worry about you, Sina. I don't want you killed or hurt again."

"I am a warrior, Salini. It is something a warrior must face."

"But this isn't your time or your fight, Sina. You can't get killed here. It isn't where you should be. Let the others do the fighting, for it is their time and fight."

"Then why are we here, Salini?"

"It was supposed to be a peaceful mission, remember? We were not supposed to have to fight."

"But there is a fight, and we are here," she argued, her voice growing more insistent. "And Salini has led the fight also, has she not? Her spear has been colored with the blood of enemies. Is it not so?"

"Yes," I sighed. "But we are not here to die."

"And yet you came with your spear, wanting to fight. Is this different from what I did?"

Well, she had me. Yes, there was a fight being forced on us, and yes we had both come prepared to be fighters in case this peaceful mission didn't turn out to be peaceful. Yes, we were warriors, leading would-be warriors in a fight not only for their islands but for their lives. And yes, we both knew that fate might have called for our deaths in any fight we engaged in. I didn't have an answer that made any sense, so I just turned my back on her and started back toward the village. I didn't get three steps before Sina's hand jerked my shoulder around to make me face her.

"We must attack them now," she blurted emphatically. "They are less now, and disheartened. This is the time to take them."

"Now?"

"It is the right time. They will not expect us to come after them so soon."

“So you want to chase them down the trail and catch them?”

“There was no fight at the beach?”

“No. It was a trick.”

"Then maybe such a trick would be well for us to use. Only we will send half of our warriors on the boat, and the other half on the trail. In that way, we will cut them in half."

It actually sounded like a good idea, though I didn’t like thinking about it. We would have to coordinate it so that our floating warriors arrived at the same time as our inland warriors did, and we had no way to communicate with each other. If either group arrived before the others were ready, we’d be fighting with only half our forces.

“If you go by land and we go by sea, how will we know when you are ready?”

“I will send a man to the top of the mountain over Tufutafoe, and he will wave his lavalava in the air to let you know that we are ready to attack. Then you will come in with your warriors by sea, and we will surround the village and come at them from the inland sides.”

So there was a way to communicate after all.

“First, I want to see how many of them fell so that I know how much weaker they are,” I told her, turning back to where the fight had taken place.

Sina’s warriors were busy finishing off the wounded, and there were several. I counted eleven men down, which reduced their numbers to around thirty or so. We had about the same number of mature fighters, plus another fifteen or so who were younger. The youngest ones were in their early teens, the ones I’d left in reserve. Did I really want to take them into a battle like this? No, I couldn’t do that. On the other hand, I might not have a choice. We’d need everybody capable of wielding a weapon if we were going to win this war.

"Did you see any wounded who escaped?"

"Yes. One man had blood on his shoulder, probably from an arrow, for my spear does not make such small wounds. Another had his hand over his side, where there was much blood. I think he will not survive. He and another stumbled much as they ran away."

"That's it? Only three?"

"Maybe more. It is not easy to see through the jungle."

"That's true," I nodded, acknowledging the truth of her statement but wishing she'd seen more. "So we have eleven dead and three injured, which means they probably lost fourteen fighters, maybe more. That still leaves about as many as we have. We have no clear advantage."

"Are you stupid, Salini? We have the warriors of Tufutafoe, who have never been defeated in battle. With them, we will have victory."

"But they aren't trained warriors like the warrior caste of your time."

"Were they trained warriors when Masana and her followers made war against the Aunu'u? No, they were not. They killed the Aunu'u because they were the warriors of Tufutafoe, and it will be those from the same village who will do so again with these cannibals, for battles are won in the heart, not the spear."

"I remember you saying that once before."

"And was it not a true saying?"

"Yes, I suppose it was by your way of thinking."

"Then we must bring together our warriors and prepare to attack. You will take half with you on the boat, and I'll take half over the trail. We will attack from all sides at the sign of the waving lavalava."

Chapter 17

Bolstered by the fact that the enemy had failed in their attack against us, our warriors were eager to commence a reprisal attack against them. They hoped to slaughter the cannibals in such numbers that it would end the war once and for all. I wasn't so optimistic, but Sina was right: a quick retaliatory strike was the logical thing to do. I took the archers and those who would man the paddles, along with another fifteen fighters armed with spears and javelins, with me. It didn't take long to get two of our boats ready to shove off, but there was no rush.

Sina and her crew had to travel over the land trail, doing so quietly to avoid alerting the enemy. That was going to take a lot longer than the boats, which could make the trip in less than half the time it took those going overland. We didn't want to arrive before the land troops, so we got everything ready and then waited. All the warriors from the Amasiala clan, led by Masotua, their chief, were going with me. They were, as Sina put it, the warriors from Tufutafoe, for they were the ones who had built the village now occupied by the cannibals. They wanted it back and were anxious to take it from the black invaders.

Those going overland gathered their weapons and supplies, held what was more of a pep rally than a war council, and set off up the trail. From previous experience, we knew it would take them at least four hours, possibly more due to the need for silence and stealth. The boats, on the other hand, could make the trip in less than two hours. Rather than bobbing around on the ocean waiting for them and risking possible detection, we chose to wait two hours before launching.

It was going to be a long two hours, because tensions seemed to grow with each passing minute. The women and children had been allowed back into the village, and it was they who started fussing and worrying over their loved ones who were either hiking overland or waiting to depart on the boats. I understood their concerns, but it was starting to erode the excitement and determination of the warriors waiting to leave. That worried me. We needed them fully committed to the task ahead, believing in a final victory, and the weeping and crying in the village wasn't helping. So, I decided that waiting for Sina and her troops was better done bobbing on the waves than letting our resolve erode away in the village.

Less than an hour after Sina and her warriors left, we put to sea, but things definitely didn't start out in our favor. The ocean was angry, with choppy waves and big swells. A wind was blowing from the north, opposing our progress in that direction. It took hard paddling to cross the reef, and even beyond it, the swells required constant effort to keep us going in the right direction. About a half mile out from the reef, which was farther than we would normally have gone in calm weather, I called a rest break for the poor men who were giving their all on the paddles.

Bobbing around was more than just a phrase that morning. The size and weight of the boat worked in our favor, but even so, we couldn't stand without being tossed overboard. We sat, being pushed slightly farther out to sea while the men manning the paddles rested. Some of the fighters took turns relieving them, but navigating those swells took a lot of hard work, and I wanted my warriors rested when the time came to fight.

The wind, which wasn't usually much of an issue on a boat of that size, was pushing us slightly southward, so despite the need to rest, we had to resume paddling periodically just to maintain our position. Once the men were rested, we started north again, fighting large swells and wind all the way. The tide was pushing us farther out, which meant we had to fight that as well. Even so, by the time we came upon the cliffs, which were the last place we could hide our approach from those at Tufutafoe, we were close to a mile from the reef.

Focusing our efforts, we turned straight toward the cliffs to get closer for when the time came to make our assault on the village. On the mountain behind the village, there was no sign of the man who would signal the time to close on the enemy, but then we were still pretty far out and Sina and her men were probably still making their way through the jungle. Normally we would have crossed inside the reef to wait, but the waves were going to make it difficult, so we were forced to bob around beyond it.

The trip had been more difficult than any we'd taken before, and I'd apparently lost track of time. Suddenly, the silhouette of a man appeared at the top of the mountain, waving his lavalava in the air furiously. That was the signal to charge the shore, but we were still far enough out that it was going to take fifteen minutes to get there, maybe longer. Sina was expecting us to come in at the same time she did, and she had no way of knowing we were so far out. If she attacked now, she'd be facing the enemy with only half our forces.

My men started paddling with everything they had to close the distance before it was too late. When we reached the reef, we still weren't in a position to see the village, but I listened desperately to hear Sina's war cry. With the lapping of the waves and the wind in my ears, though, there was no chance of hearing anything from that distance. I strained to hear anything that might indicate a battle, but to no avail.

There was no way of knowing if she was already engaged in a deadly fight, expecting us to provide support, or still waiting in hiding until we arrived. Fear seized me, and I prodded the men on the paddles to push harder, although they were already giving everything they had. Crossing the reef was more difficult than usual due to the high waves and low tide, but we finally got over it and made a beeline for the village.

Coming around the final bend before the beach opened up, we did indeed see fighting, but there was no Sina. In fact, there were no Polynesians. The blacks were fighting among themselves for some reason. The logical thing would have been to back off and let them deplete their own numbers, but by the time we saw what was happening, they had seen us. Turning back wasn't an option.

We were about a hundred yards from shore when a scream went up among the cannibals, alerting them all to our attack. Like a pack of mad dogs, they stopped fighting among themselves and started sprinting frantically toward us, spraying sand in their wake. They were maybe twenty yards from the shore at the time, so the range was too great for my archers and way too far for the javelins. Not too great for them, though. Their bows were stronger, their arrows better made, and within seconds, two of them were letting fly at us.

"Down!" I screamed at the top of my lungs, but it wasn't necessary. Everyone knew to get down on their bellies or behind the reinforcements we'd made when arrows were in the air. One of the arrows slammed into a wooden wall behind us, the other sailed over us and out to sea. We were still gliding toward shore, pushed on by the waves as if we were surfing, the distance between us closing by the second.

And still, there was no sign of Sina or her warriors.

The enemy archers shot a few more arrows, but soon realized we were protected and gave up. Seventy-five yards from shore, I motioned for my archers to shoot from a prone position, which is easier said than done. But we'd practiced it, and it paid off. Two of the enemy fighters were hit, their wounds not serious but enough to slow them down if not take them out of the fight. It wasn't really necessary for my archers to aim, since the enemy was gathered at the shoreline in a mass of bodies. A shot in that general direction was sure to hit someone.

Another volley from the archers put at least one more down, but by then, we'd closed to within about forty yards. It wouldn't be long before we bottomed out on the sandy beach, and then it would be hand-to-hand fighting. But first, we had one more weapon custom-made for that range.

"Javelins!" I shouted, and five men rose to their knees to launch our crude spears. A large man at the front of the enemy mob saw one coming at him and ducked just in time for it to take the man behind him in the chest. Another javelin went high but pierced the forehead of a man in the rear. A man at the front went down with a javelin in his stomach. We were reducing their numbers, but they still outnumbered us unless Sina showed up in time.

We hit bottom about ten yards from shore, and as a group, we jumped from the boat into the water, wading our way into the fight. The cannibals didn't hesitate, wading through the shallow water to meet us. We were just a few seconds from a close-up battle when a war cry pierced the air. A war cry I knew and loved.

Confusion broke out among the enemy forces. A few broke from their advance on us to confront the new threat, while others kept coming at us. Most, however, seemed confused, not sure which way to go or what to do. They were under attack seemingly from every direction and couldn't decide which threat was most pressing, whether to fight inland or on the shore, or not fight at all.

My warriors faced no such dilemma. They attacked fiercely, first by throwing all their remaining javelins, then going straight into battle with their spears and clubs. Most of the black fighters bolted and ran, but not to face Sina. They ran into the jungle where it sided along the shoreline adjacent to the village, and fled for their lives. In less than a minute, they'd somehow managed to penetrate jungle so thick I couldn't believe anyone could get through it, and were gone from sight and sound. The few who stayed to fight were quickly dealt with, some in a rather gruesome way. The wounded were finished off, and then we ran to help Sina.

It wasn't necessary. Her crew had killed three and wounded one, the rest had run off into the jungle like ours. Which meant we hadn't eradicated the enemy. We'd weakened them once again, but there were still some alive that would have to be dealt with. My fighters had killed eight when the wounded were finished off, and Sina's had killed four, counting the wounded man who'd been eliminated as well. That meant they were short twelve more warriors, and we hadn't suffered a single injury.

With twelve more of them dead, their total force was reduced to less than twenty. If they had any sense, they'd jump on the boat we'd left behind and go home, wherever that was. Somehow, though, we didn't think they'd do that. They might never return to Tufutafoe, where we would be sure to find them, but that didn't necessarily mean they were leaving the island. They might hide out in the jungle, or move to the other side of the island to a location we knew nothing about.

That's the trouble with the jungle. Although I love it dearly, it's a great place to get lost, to go someplace where you might never be found. We couldn't let them do that. They were a menace, and until they were eradicated or sent back across the ocean, we couldn't return to our simple village life. If the family clans went back to their individual villages, they'd be easy prey for the man-eating enemy. They could pick us off one by one from the small villages rather than facing us in a war they couldn't win. Or they might even take everyone in a village and occupy it as the Aunu'u had done so many generations ago.

We weren't going to be safe until they were completely defeated. None of them could be left alive, or permitted to remain on the island. We couldn't even allow them to occupy any of the nearby islands, which would eventually become the islands of

Samoa and Tonga. They had to go far away, or die. Those were their only options. If they hadn't been cannibals bent on our destruction, it might have been different. We might have let them build a village or two on the other side of the island as long as they didn't do anything hostile. But cannibals? Never. They had to go away or die, and we had to chase them away, kill them, or die ourselves.

Sina knew it as well as I did and took the bulk of our warriors to pursue them in the jungle, while those who remained helped to gather up the bodies, load them on a boat, and take them far out to sea. They were thrown overboard for a watery grave. Instead of eating us, the sharks would feed on them, which seemed appropriate enough to me.

When we returned, Sina was back, disgusted that they were unable to catch up with the enemy. They'd scattered in the cover of the jungle so much that it was impossible to determine which way they'd gone. Was it a preplanned escape route, or just the direction they'd gone when all other routes were covered by our fighters? Did it really matter? They'd made good their escape, but their numbers had been depleted again. There weren't that many of them left for us to fight when the time came for what we hoped would be the final showdown.

It was a good victory for us. In fact, it was nothing short of a complete rout. The enemy had suffered twelve casualties and been chased off again; we'd suffered none. It was a cause for celebration, and our fighters deserved it. The village would throw a feast and fiafia with kava and dancing in their honor, which was good. There would be speeches, songs made up and sung for them, and the women would make love to them as the heroes they were.

But the enemy was still out there. Even though there were fewer than before, we had to remain vigilant. As long as we kept the families together at Falelima, they weren't likely to attack, but until they were completely eliminated, it wouldn't be prudent for us to split up into small villages again where we would be easy pickings. Even while together, we had to send guards with our people who went out to gather food, hunt bats or birds, or fish anywhere beyond the view of the village.

With cannibals on our island, life would always be tenuous. We'd never be able to let down our guard, to travel safely, or do the normal things village life demands without an armed force going along to protect us. That wasn't something we needed to explain to The People, because everyone understood it. The cannibals couldn't be allowed to live anywhere near us, not even on another island several days' sailing time away. They couldn't be allowed to live at all.

They had to die.

Chapter 18

Sina scrounged up some supplies, gave them to two men, and instructed them to hide in the jungle above Tufutafoe and report to us immediately if the cannibals returned. The rest of us loaded onto the three boats, including the one originally lost to the enemy in our first naval attack, and returned to Falelima.

Things went pretty much back to normal after our raid. The warriors continued practicing, more weapons were made, and food was hunted and gathered just as it had been, but with an armed escort. Masana was especially helpful in that regard. Her knowledge of jungle survival skills and edible native foods was a godsend. Little Sina spent a lot of time training with the warriors, much to my chagrin, but to refuse her would have looked bad when every other person her age of sound body was training as well.

There was an incident on our fourth day back that caused mass pandemonium, especially with Sina. Another snake was encountered by a group of women gathering food in the jungle not far from the village. They were accompanied, as usual, by a team of warriors in case of an enemy threat, but the enemy they encountered was no army of cannibals. The warriors might have preferred the cannibals, though, since they shared Sina's unusual fear of snakes. To them, snakes were demons, unnatural beasts sent by Inisi, the goddess of torture and mayhem.

The whole group came screaming from the jungle, all but Masana, who had been in charge of the expedition. When I asked about her, nobody seemed to know why she hadn't come back with them, so I had one of the warriors take me to where the snake was sighted. When we got there, Masana was sitting on a fallen log, her spear resting over her knees, with a dead snake at her feet.

"I didn't realize they were so hard to kill," she said, grinning from ear to ear.

"You killed it yourself?"

"Well, I saw you kill the one in the village, and you've killed them on your archaeological digs, so I figured I could do it if you could. And I did."

It was a very young boa constrictor, maybe four feet long, and its head was smashed beyond recognition. That reminded me of the one I'd killed on my trip to Sina's time, only that one had been full-grown and a helluva lot harder to kill. In my fear and loss of composure, I'd beat its head to a pulp before Sina stopped me.

"I'm thrilled at you, Masana, but don't try taking on a big one by yourself. Any like this, though, you're welcome to. You did a good job, and it took a lot of courage. I'm proud of you."

"Thanks, Mom." There was pride in her eyes, and a smile on her face. She'd done something none of The People would do, and I knew how good that felt. I expected that The People would be even more willing to follow her now that she'd proven herself to be a fearless warrior who'd killed a "demon." She would indeed be the queen of Mutasi as far as anyone would be concerned when word of this got out.

The meal that night was meager, since the food-hunting expedition had been cut short because of the snake, and much of what had been gathered was lost in the panic that resulted. We had fish, bats, and coconuts, but no fresh produce. Nobody complained, though. The devilish snake had been killed, and Masana was the hero of the hour.

After two weeks of watching, it was decided the cannibals weren't coming back to Tufutafoe, where they'd been so soundly defeated. That meant they were somewhere else, but still a threat that had to be dealt with. Sina and I, along with a small party of men, set sail on one of the boats to run a circle all the way around the island, hoping to find their new location. It would be on the coast since living inland wasn't comfortable. Between the mosquitoes and crawling insects, the thick undergrowth, and extreme humidity, they wouldn't find life there very pleasant.

We went north, in the direction they'd taken from Tufutafoe, staying outside the reef but close enough to examine the shoreline. In places where the reef was too far out to see the shore adequately, we sailed inside it. Sailing inside the reef has its share of risks with boats like ours. The water is more shallow, so the risk of getting stuck on the coral was a valid concern. At one point, we had to get off the boat and push it over a sandbar that was so near the surface we ran aground on it. Once we were off the boat, though, the reduction in weight made it easy to push it free.

For meals, we caught fish, lobsters, and octopus to supplement the coconuts and fruit we'd brought with us or picked up during short stops. The fire pit on the boat was perfect for cooking whatever needed it. At one point, the crew speared a sea turtle with a javelin. Turtle meat isn't as tasty as fish or lobster, but at least it added a little variety.

Because we had no idea where the enemy was, we chose not to land the craft at night, preferring to sleep at sea instead. This required setting up shifts to ensure we didn't get pushed into the reef by the wind or tides. It's challenging enough to stay off the reef during the daytime when you can see the water depth and find safe places to cross when necessary. At night, the risk was substantially higher.

The first day, we passed nothing but cliffs and rocky shorelines. There didn't seem to be any beaches or locations anyone would want to live on. On the second day, as we reached the far end of the island and followed the lay of the land heading due east, we still saw nothing of the enemy or any promising spots where they might hide.

When we circled back to the southern side of the island, sandy beaches began to appear. They were small, but perhaps big enough for a small party like theirs. However, there was no sign of life on any of them before dark.

Just after we started the next morning, a javelin came hurtling towards us, falling short about twenty feet away. It was a clear sign that we had found what we were looking for. Observing it in the water, we concluded that it was one of ours, probably retrieved from wherever we'd thrown it. We never saw who threw it and found no signs of life on the nearby beach, but it seemed they were there, somewhere, likely hiding in the bush. Whether they had actually taken up residence in the jungle around the beaches, we couldn't tell from the boat, and our crew wasn't large enough to pull in and go looking for them.

We had found what we were looking for, but it didn't help. We couldn't plan an attack against a foe we couldn't see, and since they hadn't established a village, even a temporary one as far as we could tell, they could move to somewhere else at a moment's notice. No, our problem wasn't even remotely solved. They were out there, still aggressive, and there wasn't much we could do about it.

We stayed at that location for over an hour, waiting to see if they would take any further action, but it remained as quiet as the jungle can be. Nothing moved, and no dark shadows appeared that might be enemies watching us through the undergrowth. If it hadn't been for that javelin, we never would have known anyone was around and would have continued our search elsewhere. Although we were out of javelin range, we were certainly within range of their archers and took precautions accordingly, but nothing else happened.

"We must know where they are," Sina whispered quietly in my ear. That surprised me since we were likely not within hearing distance of the shore with the waves lapping against it, and even if we were, they already knew where we were, so there was no need to be quiet.

“Yes, but how?”

“We go find them. You and me. Boat goes without us.”

“You and me against probably twenty cannibals armed with clubs, bows, and our own javelins?”

“We don’t go to fight. Go to find. Then we come back with warriors, make war.”

“What you’re saying is we go farther down the coast, get off the boat, and search for them secretly?”

“Good plan, yes?”

“Except that if the boat goes on without us, how are we supposed to get back and gather our army? We’re not going overland, I can tell you that. It’s probably sixty miles from here to Falelima.”

A quizzical look clouded her features.

“That means it’s a very long distance. Too far to walk.”

“We walk far when we go to rescue Tufutafoe and Falelima, yes?”

“Yes, and it took a very long time. We’re not going to do that again. The boat will have to go farther down the coast and find a spot where it can wait for us. Some small beach will work. It shouldn’t take more than a few days to find out where they are, so the boat can wait.”

“Okay,” she nodded, “boat waits. We go.”

“We’ve got to get out of sight first, Sina. There’s no way the boat can pull in here with them watching us.”

"Boat stays here, we go. We swim." "Have you gone soft in the head? Swimming with a spear isn’t easy. We need to find some cover where the boat can pull in. Around that bend will work." "Is

Salini afraid?" There was merriment in her eyes and a childish taunt in her tone. "Yes, as a matter of fact, Salini is afraid," I replied indignantly. "Enough to be cautious, at least. And I don't like swimming with a spear." "Good! We go around bend, then swim." "We go around the bend and the boat pulls into shallow water so we don't have to swim." "Okay. Boat, shallow water, no swim. We go now." "You're sure about this?" "Are you not sure? I go myself if you not like plan." "You would, too. Wouldn't you?" "Of course." "Why?" "I am a warrior. It is what a warrior does." "And I'm not a warrior?" "Yes, Salini great warrior. That is why she comes too." "No, I go to keep you from getting killed, that's why I go. Sometimes you jump dangerously fast into things that get you in trouble." "I no jump, I go. We go. Is good plan, yes?" "Whatever," I shrugged, then instructed the crew to continue down the coast. I'd tell them to pull in when I found a good spot. Until then, they could think we were just continuing our tour of the island.

I had to admit, without the businesses and hotels and tourists that littered the place the first time I came here, the island that would eventually be known as Savai'i was even more beautiful than it would be in my time, and it was gorgeous then. The stunning canopy of the rainforest, with the lush undergrowth that was a lot easier to look at than walk through, dotted with bright and colorful wildflowers of a thousand species, made it like a heaven on earth. I tried to focus on that as we made our journey around the bend.

Once we were well beyond where the javelin had been thrown at us, out of sight and not near any other beaches that might harbor the enemy, Sina and I slipped off the boat into relatively shallow water. We left instructions for the men on board to pull into the next beach unless there were indications that it was inhabited. There they were to wait for us, and it might take several days to find the cannibals and get back to them. Under no circumstances were they to engage the enemy. If they encountered them, they were to put to sea again and get away.

The boat couldn't get close to shore where we got off, and although the water was only chest-deep when we started, it got a lot deeper before we reached the shore. So despite the challenges our spears created, we ended up swimming after all. It isn't easy with a weapon too heavy to float, sort of like trying to swim with an iron rifle. It was about a hundred and fifty yards to the shore in water that turned out to be ten to fifteen feet deep in places. There were huge coral spires that came almost to the surface, as colorful as the flowers on land, but we had to avoid them because of the dangers they represented. Some types of coral have sharp jagged edges and can cut deep gashes wherever you touch them. It took us a good half hour to finally reach the shoreline, which was rocky and overgrown, with jungle growing almost to the shore.

Because we didn't know if the enemy was nearby, we had to be quiet, but trying to crawl through jungle undergrowth on a warm dry day isn't usually a quiet enterprise. It's thick and tough, with some plants that cling to your skin or clothing, so the going was slow. We listened to every sound, even the birds, alert for anything suggesting that others might be nearby, but all seemed perfectly normal. Once we reached the slope of the mountain, we went up it a fair distance before cutting to the left. That pointed us in the direction from which the javelin had been thrown. That way we'd be above anyone on the shoreline, which is where we expected to find them.

It took hours to reach the small beach from which the javelin had been thrown, but from our hiding place, it looked deserted. For a long time, we watched for movement or any sign of life without finding anything. We listened for any out-of-the-ordinary sound, but everything was normal. The birds and insects continued their sounds uninterrupted, and except for a few birds that flew over the beach, nothing moved. There were no footprints in the sand that could be seen from where we watched.

Whoever had thrown that javelin must have been a lone hunter or fisherman, taken by surprise when he saw us. If nothing else, it suggested that the cannibals weren't too far away, but didn't pinpoint where. It was still a guessing game. If we could capture that hunter, we might be able to force him to tell us where his people were, but it appeared he was gone. Unless we could track him back to his group, which isn't easy in a jungle like that, we still had no idea where the enemy encampment was.

We were just starting to rise from our hiding spot to go looking for tracks that might help us in our search, assuming he was gone, and then we saw him.

Chapter 19

He was sitting in the shade of a huge banyan tree, leaning against the trunk where he could watch the shoreline without being easily seen. We dropped back into hiding, both of us surprised at how close we'd come to giving ourselves away. He appeared to be alone. If he was watching to see if our boat returned, he didn't seem overly concerned. In fact, he might have even fallen asleep. We couldn't be certain, and approaching without being heard would be challenging if he were awake, but capturing him was crucial. He needed to be taken alive to tell us where the others were, though there was always the possibility that more of them were hidden in the nearby jungle.

Was he a lone lookout, watching for threats to his people? If so, why hadn't he gone to report that our boat had come by? It didn't add up, but we had no way of knowing how cannibals think. Did he believe he could intimidate us away by himself with a crude javelin, or confront us in single combat? Perhaps the fact that our boat had continued down the coast had given him a false sense of security, leading him to believe we hadn't discovered where he and his people were hidden.

Or perhaps it was something else entirely, something we were overlooking as we watched him. There were too many questions and not enough answers. He seemed too relaxed to be a lookout, but what else could he be? A lone hunter, perhaps, catching bats and birds, or gathering edible plants?

Sina gestured for me to stay put. She pulled out her knife and began creeping toward the beach where the lone enemy sat, moving slowly and quietly, spear in her left hand and knife in her right. It reminded me of the time she'd instructed me to stay by a kapok tree while I ended up accidentally killing a war chief of the Aunu'u on my first trip into the past. Back then, I was a novice, barely starting to learn the ways of a warrior, and could have easily been the one who ended up dead. Was I about to face another enemy now? Hopefully not, but this time I was not only trained; I had gained substantial experience. I was better prepared to face an enemy now than I had been back then, and my spear was superior.

I kept a close eye on the surrounding jungle, ensuring no one ambushed Sina as she made her stealthy approach toward the lone man. If anyone else appeared, it would likely mean the whole tribe was nearby, and I'd have to intervene—just the two of us against twenty or more. The odds were not in our favor, even for trained warriors with superior weapons. Fortunately, nothing stirred. The birds continued their songs, the insects didn't falter in their chirping, and there was no sign that anyone else was around.

Then, when Sina was about thirty feet from him, another man emerged from the jungle to my right. He moved quietly, stalking not Sina, but the seated man, and hadn't noticed Sina approaching. She saw him, though, and quickly ducked out of sight.

The new arrival was tall, lithe, muscular, and his skin was dark as the night. His hair was short and curly, he wore only a brief loincloth, and he held a war club at the ready. The seated man hadn't noticed him, likely asleep or close to it. Suddenly, the hunter sprinted the last few feet and struck a deadly blow to the seated man's head. It wasn't a fight; it was an assassination. Why would he kill one of his own?

As the assailant examined his handiwork, Sina sneaked up behind him and pressed her knife against his throat. He started to turn, ready to fight back, but she stabbed him in the stomach before he could. Dropping his club, he opened his mouth to scream, but Sina quickly clamped her hand over it, pushing him to the ground to muffle his cries.

For a moment, I stayed back, scanning for any additional threats, but the area remained still. I then moved in to assist her. She whispered fiercely into his ear, one hand still over his mouth, the other holding her knife at his throat, but he resisted. He tried to sit up; she pushed him back down forcefully. When I reached them, she had me hold my knife to his throat. Removing her hand from his mouth, she dropped her knife to his loincloth. With the tip, she lifted it, revealing his genitals, and rested her blade against his skin. He ceased struggling immediately.

"I will cut this off," she hissed softly, "if you do not tell me where the rest of your warriors are."

He was bleeding badly from the stomach wound, likely facing a slow, painful death. Yet, the immediate threat to his manhood widened his eyes with terror. He made a feeble attempt to struggle, but Sina's blade drew a bead of blood from the base of his penis, causing him to freeze.

"Tell me, or I make you a girl," she threatened, slightly pulling her hand from his mouth but ready to clamp it shut again.

"They are at Atu," he whimpered.

"Atu? Where is that?"

He jerked his head toward the direction he had come from.

"How far?"

"Many stone throws. Maybe half a sun walk."

"Why did you kill this man?" she asked, nodding toward the man with a bashed-in head.

"He deserted, said he would no longer fight."

"So you killed him. Does that make him a better fighter?"

"It scares others from leaving. Many are not happy, want to go away. I kill any who try to run, like this one, so no more will run."

"Did any of the others come with you to kill him?"

"I came alone. Didn't need others for this."

"How many are left?"

"Not so many as before. The war goddesses have great power."

"But how many?"

"This, I do not know," he said, extending his hand and flashing his fingers five times, likely indicating twenty-five. "Maybe so many as that, I do not know."

"Would the rest leave these islands if they had a boat?"

"No. Aiatu says no. We fight, we kill, we eat."

"Who is Aiatu? Your chief?"

He nodded. "He has great power. Wants to fight, kill, eat. Gain the power of war goddesses."

"How would he do that?"

"He eats, gains power."

"So, he wants to eat me and gain my power?"

"Yes."

"Were you going to take this man back and eat him to gain his power?"

"Yes, as much as I can carry, but he had little power."

"Because he was a coward?"

"Yes."

"And if I kill you, will you have power?"

"If you eat me, you get my power."

"Do you think you would taste good?"

"Yes, it would be so."

"You are dying, you know. You have lost much blood."

"It is fitting for a warrior to die so."

“You are no fitting warrior, who preys on the weak of his own people while they sleep. No, there is no power in you.”

“I am a warrior with great power,” he objected, his pride wounded.

“I promise you, you will die as what you are, warrior or assassin.” With that, she dropped his penis and brought her knife to his throat, slicing it to the bone. He bled profusely, convulsed briefly, and then lay still, just another corpse on the sandy beach.

“Well, we found what we were looking for,” I pointed out, “so now what?”

“Atu is not found. We floated past all sands in the direction he showed, and saw no enemy. It is unknown to us and must be found.”

“Yes, but it’s a half day’s walk from here.”

“Then we walk half a day.”

In reality, it took well over half a day due to the need for caution and silence, navigating through dense undergrowth. The man she had killed must have used some kind of trail, but we found nothing resembling one. It was jungle and thick timber all the way. Twice we encountered beaches we had passed by boat, but found no signs of human presence. We finally stumbled upon them by sheer accident.

The sound of voices alerted us. By then, we were too close and had to retreat to a safer distance. They weren’t camped on a beach, which is why we had failed to see them from the boat. Instead, their camp was at the base of a lava rock slide. It looked uncomfortable, but it effectively concealed them from ocean view, where any searcher would likely be. No search from a boat, even in a small paopao, would find them; from that perspective, they had chosen an excellent spot.

The piles of leaves, ubiquitous in the jungle, were arranged in hollows and crevices in the lava, serving as makeshift sleeping mats. These were rugged men, evidently unafraid of enduring the wilderness. Better that than being discovered by an enemy that had proven more formidable than they had anticipated. They were down to a fraction of their original strength, having made little to no progress toward achieving their goals.

They did not appear to be a content group. Their conversations were loud and harsh. There was no laughter, no camaraderie, only grim faces and terse words. I thought I saw blood on the cheek of one man who stood apart from the rest, but it blended into his dark skin, making it hard to be certain. Another man limped painfully, likely wounded in a previous battle. At the bottom of the slope, another man hung by his feet from a tree limb, obviously dead and partially consumed.

Most had war clubs, some crudely fashioned from tree branches, likely replacements for more effective weapons lost in battle. A few held javelins left by our warriors or ones they had crafted themselves. The proximity of weapons indicated a constant readiness for conflict, suggesting we might not be the only threat they feared. This was far from a unified army.

Sina began to crawl down the slope toward the shore, and I followed silently. We spoke only after reaching the coast, away from their camp, where a rocky cliff dropped twelve feet into the waves below. She studied it carefully, memorizing each detail so we could find it again when we returned with enough force to finish what we started. That cliff would serve as our marker, indicating that the enemy lay directly inland.

"They are no more warriors," she began, "they are frightened men. There will be no glory in killing them."

"There is no glory in killing anyone," I retorted, "though sometimes it must be done."

"Yes, it must be done. They eat their own men raw. Did you see?"

"Yes, I saw. It's horrific to think of such a thing."

"These men do not deserve to live, nor do they deserve to die honorably. They should be squashed like centipedes."

"What's the plan for taking them?"

"I will think on this. You should think also, for you make good plans. It will not be easy for our warriors to sneak up on them, for the jungle will make much noise. Or it may make no noise."

I knew what she meant. The birds and insects of the jungle, once they get accustomed to the presence of humans, resume their normal sounds. However, a new group moving through the jungle will cause them to go silent or fly away. Such silence can be as revealing to an astute enemy as the sound of men crashing through the brush. Either way, their chosen spot made it difficult for us to approach stealthily.

"We go to Falelima, make a plan to kill these things who think they are men."

"We need to find the boat, because it's a long walk all the way around the island."

"Is Salini, the warrior from Siata and adopted daughter of Tufutafoe, afraid to walk?" she smiled.

"No, Salini is not afraid to walk, but she prefers the boat."

We made it back to the beach where the two dead men still lay before dusk. We dragged the bodies to the ocean and dumped them, hoping the tide would carry them farther out. Then, we spent the night on the soft sand. At first light, after a breakfast of coconut juice, we set out in search of the boat, which turned out to be farther than we expected. It took us until mid-afternoon to locate it. The crew gave us a hearty meal, and then we pushed off, back on the ocean, debating whether it would be better to return the way we'd come, or to continue the way we had been heading. Were we halfway around the island, more than halfway, or less?

Ultimately, we decided to press on and explore the parts of the island we hadn't yet seen. It turned out we were a long way from being halfway. Four days later, we finally arrived back at Falelima, but things had happened in our absence—things that were going to have a major impact on our plans for yet another war.

Chapter 20

Everything looked quiet as we came into sight of Falelima, but as we pulled in, a host of villagers rushed out to greet us. Apu, Little Sina, and Masana waded out waist-deep, then had to backtrack as the boat nudged by them on its way to the shore. We received a warm reception, but their faces bore an aura of trouble. Aside from the welcoming party, the village looked eerily empty.

I stepped off the boat into knee-deep water, where Little Sina met me with a big hug. Apu followed with a hug as well, and then Masana took her turn. They all greeted Sina in a similar manner. It felt good to be back among them, to feel their love and know they were safe after our long absence.

"There is a problem," Apu stated emphatically. "The People, they go."

"What do you mean, they go?"

"They say, 'bad men go that way, our village this way, we go home.'"

"How many?"

"All. Only Falelima people now in Falelima."

"Everybody left?"

"Yes. It is so."

Sina joined me, having overheard. "This mission of peace, not much peaceful. These people are foolish. I think I go back to my people."

"If only we could," I groaned. "Now what do we do?"

"We go home. These people are foolish, and I no longer fight for them."

“But the Little People won’t take us home until it's finished, so how are we supposed to get there? We can’t paddle a boat to the past or future, and it’s a very long walk.”

“Little People are foolish, too.”

“I won’t argue that, Sina, but it changes nothing. We must finish what they think our mission is before we can go home, so we have to make a plan to get it done. What do we do next?”

“We kill cannibals. All cannibals.”

“Sure, but how? We’ve lost our army, all but four or five who might not be willing to fight against such odds.”

“Then we bring the army back.”

That was the only feasible answer, of course, and we both knew it. It meant a long delay, during which the enemy might move its camp. That would mean starting all over again with a search for their location, and The People were so anxious to return to their regular, tribal lives that they might not have patience for another excursion.

“When did they leave?” I asked Apu.

“Two, three days ago.”

That meant they’d all reached their villages and started settling in. For the moment, they were safe because the enemy was far to the north and they were to the south and east, but that could change. The smaller villages would make a much more attractive target than Falelima had been with all The People gathered as a united front.

Petaio, the chief of the Falelima clan, came to greet us as we waded out of the water, his face beaming. “It is good you have come back,” he began. “Your fale is empty and anxious for you to rest in it.”

“It won’t be empty long,” I told him. “We’re going to bring the others back.”

He seemed genuinely surprised, as if our return meant the threat had been neutralized. “Is it so?”

“It must be so,” Sina interjected. “The cannibals are not dead, and if we let them live, they will kill us. The People must come back.”

That wasn’t what Petaio wanted to hear. Hosting all the families in his small village had been a strain on his hospitality, using up resources his own family would need. The search for food had taken them much deeper into the forest and along the coast than would normally be the case, and it was going to take time for those resources to renew themselves. The masi pit was ready now, but even the most hungry avoided it except as a last resort to stave off starvation.

I didn’t blame them. The food in the pit had fermented to a soft, cheese-like state, but the smell was so strong and foul no one wanted to go within a hundred yards of it. It was clear why they’d dug it so far from the village. Despite the smell, it didn’t taste too bad if you were hungry enough.

“Maybe you just bring the warriors back,” Petaio suggested.

“That would leave the women and children alone and helpless,” I argued. “The cannibals would love that.”

“But they are far away, are they not?”

“Not far enough,” I explained. “By land, they could be here in two days, or they could build boats and be here in one. No, they must be killed or driven off the island for us to be safe.”

That was the bottom line. The enemy would never stop their assault on what they saw as easy targets, which is in reality what we were but couldn’t let them know. We had to attack and decisively destroy them for the safety of our villages. There could be no other option.

"We'll rest for a few days, then seek out the other villages," I told Sina.

"Today we rest," she replied, "tomorrow we go."

It felt good to be back with my family, but I caught Sina casting a wistful glance my way, a reminder that her own family was not with her. I understood her eagerness to conclude this mission and return to them, and I shared her desire to return to my own time as well. This mission hadn't captured my heart as deeply as the previous ones had. Despite my affection for the future Samoans and their islands, I harbored no thoughts of remaining indefinitely, unlike during my accidental first journey into the past.

The night was somber. Apu and I sat on the malae, gazing up at the myriad stars, a sight never seen in Boise due to the light pollution. The stars looked different here in the south. I've never been good with constellations, barely distinguishing planets from stars, but I've always been captivated by starry nights. They speak of infinity, of realities beyond our immediate presence, transcending any news or historical events. They are ancient, a billion times older than any story ever told.

We retired to our fale late, planning to sleep in the next morning. However, our plans were interrupted by a cry from a boat crossing the reef just as dawn's first grey light touched the eastern sky. It was a fishing paopao, and the man aboard was urgently shouting, despite the early hour.

"Queen Masana," he called out while still fifty yards from shore, "Queen Masana, I bring news. Very bad news. Queen Masana, I must speak to you."

Masana hurried out in response, clad only in her lavalava. Being topless is common among The People, though I hadn't seen Masana like this before. Little Sina hadn't adopted the local custom either, which I appreciated, though I wouldn't have minded much; after all, during my unintended trip to the past, I had found myself similarly undressed.

"What do you want?" Masana called out irritably, disrupted from sleep. The man beached his boat swiftly and ran up the malae to her.

"A bad thing happens," he panted. "Many boats come. Many dark people who look like cannibals. Men, women, children. They all come."

Masana instantly snapped to alertness. "How many?"

"Many. As many as the stars in the sky. Many boats."

As I approached, Sina was right behind me. "What is it?" she asked.

"I think he's saying more cannibals are coming," I relayed, hoping I had misunderstood.

"How many more?"

"A lot, according to him. And they're bringing their families."

"How many boats?"

"He couldn't say exactly, but it sounded like many."

"This is not good, Salini. How can we fight so many?"

"It is not good," I agreed. "The first thing we need to do is send a messenger to bring all the villages back here. We'll try to come up with a plan while we wait for them."

"What plan can we make? There are too many of them. It is my thought that we must leave this island to them, and go to another."

"No!" My response came out more forcefully than I intended. "This is the island of The People. No one will drive them away from here. We will find a way to deal with this."

"You have gone soft in the head, Salini. To fight is good when there is a chance to win, but not when it means certain death."

"Is this the word of Sina, war chief of Tufutafoe and defender of the forest?"

That angered her, which was precisely my intention. "I am not afraid to die, Salini," she snapped, "but I do not wish to make my people die with me when the enemy is too strong to fight. If you have a plan to save us from this menace, say it now."

"There might be a way," I sighed, knowing it was a long shot. "How many warriors do we have in the village now?"

"Six. Maybe a few more, but it is not enough."

"Okay, the first thing we need to do, in addition to getting the other villages back here, is to find out how many boats they have, where they're planning to land, and how long before they arrive. If they're coming here, and it seems they probably are—this is where they think their warriors are—then we must meet them with a weapon they are not expecting us to have."

"What could such a weapon be?"

"Let's send a messenger to the other villages, and dispatch a boat to gather intelligence on the approaching enemy. Then you and I can devise a plan based on what we learn," I suggested.

Just then, Akime came running up to Masana, gasping for breath from his sprint from the far end of the village. "What is it, Masana?" he asked urgently.

"More enemies are coming," she informed him. He was momentarily stunned, but quickly composed himself.

"You need not worry," he assured her confidently. "I will protect you."

"Thank you," she responded, her tone thoughtful yet subtly dismissive of his overture. Masana was undoubtedly more battle-hardened than Akime, yet she chose not to belittle his offer of protection. I found her restraint admirable and wondered if there was more between them than I had realized. Though I hadn't had much personal contact with Akime, it was clear he was often by Masana's side, seemingly eager to do anything for her. The young man was clearly smitten with her, but did she reciprocate his feelings? It was beginning to look like she might.

Within an hour, Matapusi was equipped with supplies and sent off in his fishing paopao to rally the other villages back to Falelima. Simultaneously, Atepiso set out in his own paopao to scout the ocean and assess the incoming threat. His mission was to report back with details about their location, numbers, and likely destination, providing us with critical data to formulate our defense strategy.

I shared Sina's sentiment; fighting an overwhelming force seemed futile, so we needed to find a strategic alternative. Victory must be achieved without a bloodbath, if at all possible. Everything hinged on the timely arrival of the newcomers, their numbers, and their intended landing site. I suspected they aimed for Falelima, where the initial group had presumably scouted in advance for the larger contingent, although this was still conjecture.

Hopefully, Atepiso's reconnaissance would yield the necessary insights. We were desperate for information: How much time did we have to prepare? How many warriors were we up against? Where exactly would we confront them upon their arrival? The answers to these questions were crucial and hopefully imminent. And critically, even counting our young reservists, we did not have enough warriors to face such a vast horde.

The stark reality was clear: we had to devise a method not only to survive but to triumph in a war that had been thrust upon us, despite being massively outnumbered.

Chapter 21

It was good to be back with my family again, but I noticed Sina giving me a wistful look, a reminder that her own family wasn't with her. I understood her eagerness to conclude this mission so she could return to them, and I shared her desire to return to my own time as well. This mission hadn't captured my heart as deeply as the previous ones had. Although I had grown to love the future Samoans and their islands, I harbored no thoughts of staying forever, unlike during my first accidental journey into the past.

The night was somber. Apu and I sat on the malae, staring up at the myriad stars, a sight unseen in Boise due to light pollution. The stars looked different here in the south. I've never been good with constellations, barely distinguishing planets from stars, but I've always been enamored by starry nights. They speak of infinity, of realities beyond our immediate existence, transcending any news or historical events. They represent ancient, timeless vastness.

We retired to our fale late, planning to sleep in the next morning. However, our plans were abruptly changed by a shout from a boat crossing the reef just as dawn's first grey light touched the eastern sky. It was a fishing paopao, and the man aboard was urgently shouting, despite the early hour.

"Queen Masana," he called out while still fifty yards from shore. "Queen Masana, I bring news. Very bad news. Queen Masana, I must speak to you."

Masana hurried out in response, clad only in her lavalava. Being topless is common among The People, though I hadn't seen Masana like this before. Little Sina hadn't adopted the local custom either, which I appreciated, though it wouldn't have bothered me much if she had. After all, during my unintended trip to the past, I had found myself similarly undressed.

"What do you want?" Masana shouted back at him, clearly irritated at being woken so early. The man beached his boat swiftly and ran up the malae to her.

"A bad thing happens," he panted. "Many boats come. Many dark people who look like cannibals. Men, women, children. They all come."

Masana instantly snapped to alertness. "How many?"

"Many. As many as the stars in the sky. Many boats."

As I approached, Sina was right behind me. "What is it?" she asked.

"I think he's saying more cannibals are coming," I relayed, hoping I had misunderstood.

"How many more?"

"A lot, according to him. And they're bringing their families."

"How many boats?"

"He couldn't say exactly, but it sounded like many."

"This is not good, Salini. How can we fight so many?"

"It is not good," I agreed. "The first thing we need to do is send a messenger to bring all the villages back here. We'll try to come up with a plan while we wait for them."

"What plan can we make? There are too many of them. It is my thought that we must leave this island to them and go to another."

"No!" My response came out more forcefully than I intended. "This is the island of The People. No one will drive them away from here. We will find a way to deal with this."

"You have gone soft in the head, Salini. To fight is good when there is a chance to win, but not when it means certain death."

"Is this the word of Sina, war chief of Tufutafoe and defender of the forest?"

That angered her, which was precisely my intention. "I am not afraid to die, Salini," she snapped, "but I do not wish to make my people die with me when the enemy is too strong to fight. If you have a plan to save us from this menace, say it now."

"There might be a way," I sighed, knowing it was a long shot. "How many warriors can we gather if we include everyone? Even those who were to stay in reserve?"

"The other villages won't be here for days," she replied, as if I was too foolish to know that.

"How many now? Not when the other villages get here, but now?" She held up both hands and spread her fingers to indicate ten. "What if we include the younger ones and the strong women?"

"Maybe fifteen or twenty, but it isn't enough to fight so many."

"Forget those on the boats, Sina. What if we could capture the ones already on the island? Then we could use them to barter with the new arrivals. If they want our captives alive, they have to leave the island and go back where they came from. Do you think that might work?"

"It will not work, for we have killed so many. There was maybe a hundred, but now there are twenty, maybe a little more. I think they will be angry and will not do as you say."

"If that be so, there will be twenty fewer warriors we must fight."

She nodded her head gravely at that, considering it carefully. "It may be so, but we have no time to capture them. It is said the boats will be here tomorrow."

"Then we must capture them tonight, while they sleep. We can make our ten or fifteen warriors sound like a hundred if they act properly. Fear of such numbers, or what they think our numbers are, will make them surrender."

"Tonight? Are you soft in the head? How can we attack them tonight, when they are camped in the jungle?"

"We have the advantage of knowing where they are, and can get there from Tufutafoe, even in the dark. If we approach quietly and have our people surround them, then let out war cries from every direction, they will think we are many and surrender. That is my hope, at least. Then we bring them here to await the arrival of their boats, displaying them openly on the malae so those on the arriving boats can see them. Then we offer to let them live if their people will go back where they came from. It may be that their wives or children will be on those boats, and will want them spared. Perhaps enough to do as we say. If they don't agree, we kill the captives while they watch."

"But what if they do not agree? They are many and we are few."

"Then we have no choice but to make war and kill them, although I hope it doesn't come to that. We have one advantage, though. We have a fairly good idea how many they are, but they have no guess as to how many we are."

"Does it matter? They are more than we. If they choose to fight, can we face them with a few warriors, mostly women and children?"

I laughed bitterly. "We will need to make them think we are many, just as we will do to those we attack tonight." A memory struck me, and I grinned as I went on. "But even if it was only you and me, Sina, I would take up my spear and fight, for a battle is won in the heart, not the spear. Do you not remember those words, spoken by the war chief of Tufutafoe and defender of the forest long ago?"

"Do you mock me, Salini?"

"No, I do not mock you. I merely repeat the wisdom of a warrior spoken in another time, facing another battle. Do you not have the heart for a battle if it cannot be avoided now, in this time? Capturing those who are on the island now may be a way to fight a small battle in order to avoid a big one."

“I think it is Salini who has the heart of a warrior,” she sighed. “Do you truly think this will work? Is it your true thought that those warriors will surrender?”

“If they think they are surrounded by many, especially when it is dark, why would they not? I think they do not wish to die any more than we do. Some of them, after seeing so many of their own who have died, are tired of fighting and want no more. We saw that in the man who was killed at Tufutafoe, did we not? He was killed by his own people because he wanted to fight no more. Maybe there are more of the same mind. Perhaps even many.”

“Perhaps,” Sina said with a doubtful sigh, “but it is a plan that may result in the deaths of many of our warriors.”

“That is possible,” I admitted, “and if you have a plan that does not involve such risks, I will hear it. If we do nothing, and the enemy boats arrive with more warriors, all of The People will likely die. Is it not better to risk a few deaths than to wait for all to die?”

Masana wandered over just then, followed by Little Sina and Apu. Akime followed Masana, but stopped farther back, standing quietly as if he didn’t dare join our group. He was a good looking kid, intelligent and shy, but not afraid. I couldn’t help wondering if there was a mutual feeling between them, having seen them together so much. Akime had vowed to protect her against harm, which is a vow that takes courage to keep. Masana was engaged to a man in my time who’d refused to come with her on this quest. A man who wouldn’t defend and support his woman, so it would be no surprise if her attentions turned to someone else.

“What’s up, mom?” Little Sina asked, noting the sour look on Sina’s face.

“Salini has a plan,” Sina replied unhappily. “It is a bad plan that I do not like.”

“That’s what’s up,” I told Little Sina. “We need to take action before the numbers turn against us even worse, but haven’t come up with anything yet.”

Sina shook her head in response to that. “The plan of Salini is a bad one that I do not like, but I have no better plan. It is in my mind that we should do as your mother has said.”

“So what’s the plan?” Masana threw in.

“We will take all who can fight and capture the enemies in the jungle while they sleep,” Sina explained, “even though they are more than us. We will capture them alive and bargain with those on the boats for their lives. Either they leave the island and go back to their own lands, or we kill the captives and make war with those who are coming.”

“That sounds like a good plan,” Masana said, showing her white teeth in a wide smile. “We go as soon as the other villages arrive?”

“We can’t wait for them,” I said, butting in. “The boats will probably arrive tomorrow, so we have to make our move tonight. I don’t think those in the jungle know about the boats, which is why they’re hiding rather than attacking. They have no idea support is so close, so they’re more vulnerable than they would be once they know about the boats.”

“So we’re going to go after them with just the fighters we have here?” Masana was as skeptical about that as Sina had been.

“Yes!” Sina declared with unexpected bravado. “We go with those who are here, for they are like the warriors of Tufutafoe in olden days, who have never been defeated. The warriors of Falelima will prove themselves as brave and fierce as those of Tufutafoe."

"But they are not as skilled or experienced as the warriors of old," Masana pointed out, using an argument that had been gone over a dozen times, if not more.

"Tonight they will be," Sina replied. "Were those of your time skilled and brave, Masana? No, I say they were not, yet they defeated the Aunu'u. We will do so again with this enemy, for we are no less than those of your time or mine, and we will surely die if we don't succeed."

"I will fight," Apu broke in, "and I will be as strong and brave as those of Tufutafoe in my time, though I be but a man."

"And I as well," Masana promised.

"I fight too," Akime said, speaking out loud for the first time I'd ever seen.

"Me too," Little Sina piped up to my chagrin. I didn't want her anywhere near a battle, but knew I couldn't refuse to let her fight. Sina would never forgive me for that.

"Tonight," Sina nodded, "all will be warriors, brave and true. And we will capture the enemy. They must be taken alive, so kill only if it must be. Masana, gather everyone who can fight and bring them here, for we have much to do."

Masana took off to do what Sina had said, followed closely by Akime. Less than an hour later we had eighteen men and women, some still in their teens, who were eager to join us in what seemed like a last stand campaign against the enemy. Sina started laying out a battle plan with the group that included Little Sina, which bothered me. I wanted her next to me when the fighting started, but it would be dark, which would make it difficult, if not impossible, to protect anybody.

It was a big enough group that one boat would be sufficient for us. Our hope, however, was to bring back around twenty captives, which meant we'd need two boats. The group of rag-tag warriors was split into two groups, each with enough strong men to man the paddles. It was important not to work them too hard on the paddles, for we'd need them to be capable of fighting when the time came. It was going to be a long night for everyone, so exhaustion could be as much of an enemy as the black fighters.

Sina did a final check of everyone's weapons to ensure there were no cracks or breaks that would leave them weak when needed. She ran a few drills and even spent some time making them practice their war cries. The evening meal was served, ensuring all the combatants were well-nourished, and then an impromptu cheer session was held in honor of those about to go to war. There was some dancing, speeches, and kava. To my surprise, a few of the dances were actual war dances, simulations of the battles we'd already had with the cannibals, performed without practice but with a level of energy not unusual among The People. They ended with a vision of how tonight's battle would go, predicting a victorious outcome. I could only hope they were right.

Being the brainchild behind the battle plan was anything but comforting. I was nervous about all the things that could go wrong and scared for our fighters. If we were detected by enemy observers before we got in position, things could rapidly turn against us. If we failed to make them think there were more of us than we really had, the same would happen. And there was always the chance that the enemy had moved to a new hideout and we wouldn't find them at all. It would take perfect coordination, good timing, and a lot of luck for things to turn out the way we wanted, with many variables that could make them go the other way.

On the other hand, my knowledge of the future helped to ease those concerns a bit, but not completely. I knew that in my time The People would live on this island, as well as many of the others that were close by. This would be Samoa then, and this island would still be called Savai'i, the name it had already been given. Logically, that meant that The People had to overcome this obstacle. At least, that's what I told myself, but it wasn't necessarily a given. The People could lose now, and others could arrive sometime later to reclaim the island and take it back from the cannibals. If that was the case, I might be called back by the little creatures to fight yet another war, provided I survived this one.

What would happen if I died in this time? Worse, what would happen if Little Sina or Apu died? My family, in my own time, would simply disappear. There would probably be a police investigation, extended searches to find us or our bodies, but nothing would be found. We just wouldn't be there, with no explanation as to what happened to us. Or, if at least one of us returned to the future, his or her explanation about the disappearance of the others would be so unbelievable that it would probably result in murder charges being filed.

No, we had to survive this. Me, Apu, Little Sina, Masana, and Sina. And The People. That's all there was to it. We had to survive and keep this island for those who would occupy it for thousands of years. We couldn't leave it up to a future migration of The People, we had to do it now. By ourselves. Regardless of the odds against us, we had to win a victory here and now.

Those who waded onto the boat with us didn't think about the possibility of other groups coming later to reclaim this island. Even if they did, it wouldn't have mattered to them. Their lives were on the line, and the lives of their families. To them, this wasn't a moment in history, it was a life or death encounter in the now. They would defeat the cannibals or die. It was that simple. Anything beyond that was meaningless.

We put to sea just after dark, with no torches to guide our way that might be sighted by the enemy. We were going to navigate our way to the coast of Tufutafoe with nothing but the light of a quarter moon and the stars. Even that would be blotted out when we entered the forest, but everything depended on us getting to the encampment of the enemy quietly and without being detected.

As difficult as that would be, it was only the first step in what we needed to have happen. Once we surrounded the enemy without being detected, our band of young and inexperienced warriors, though outnumbered, would have to make the cannibals surrender. We had to make them think our numbers were so great, and their chances of escape or survival so small, that they would give up without a fight. It was a lot to hope for, but our survival most likely depended on it.

I was heartsick as we made our way along the reef on our way to an encounter that would have repercussions that would extend all the way to my time, thousands of years in the future of those who were with us. Maybe they didn't think about it in those terms, but I certainly did. Little Sina and Apu were tense, afraid of the coming battle and rightfully so, but no one was more worried or afraid than me.

Chapter 22

A storm moved in. Not a big one, but with enough clouds to blot out the moon and stars, which The People are so good at navigating by. A tropical rain dumped on us, which was nothing new or unusual. In that climate, the rain isn't cold, and being wet isn't really a problem, but losing sight of the lights in the sky was. In the dark, we couldn't see the coast either, and knew where the reef was only by the sound of waves crashing against it.

It certainly wasn't the start we'd hoped for on this crucial mission. We gauged where we were mostly by how long we'd been out, which isn't very reliable. Despite the storm, there was little wind and the ocean was relatively calm, so we figured we must be making pretty good time. A dim grey line of whitecaps could be seen where the waves formed on the reef, another line even dimmer along the shore.

Any other time the rain would have been welcome, but when you're trying to see the coast on a black night, the rain running off your forehead into your eyes doesn't help. Little Sina, less accustomed to it than the rest of us, kept gasping as if she were drowning. Tropical storms can make you feel that way. They can drop six or more inches of rain in a matter of minutes, almost like being under a waterfall you can't move away from.

Little Sina, like everyone, was afraid of the upcoming battle but tried to hide it. I was afraid for her more than for myself, and didn't try to hide that, either. Apu, seated next to her, kept running his hands over his spear. It wasn't an act of care for the weapon or endearment for what it could do, but tension. We were all pretty uptight, especially given the way the weather had turned on us. Some thought it was a bad omen, but Sina silenced any such talk.

The coastline and sea blended into each other. All we could see were the misty white lines where waves crashed against the shoreline. Because it was low tide, waves formed on the reef, making it more difficult for us to know exactly where we were in relation to the coast. The bigger waves on the shoreline were those crashing against rocks. Those lapping onto a sandy beach were smaller. Too small to see? We hoped not, because that might be the only indication of where Tufutafoe was on a night like this. The waves on the reef weren't big, but they helped to show where the water was too shallow to cross.

Once we hit something, bottoming out on coral. We'd hit a shallow reef without realizing how close we were, which can be dangerous in rougher seas. The men were able to paddle us off without a problem, but from then on, we had to try to stay in deeper water farther from the reef. Which put us farther from what we needed to see. We moved ahead slowly, searching the blackness for any sign that might tell us where we were, or better yet, where Tufutafoe was.

The rain stopped suddenly, as if someone had turned off a switch. Rains in the tropics are often that way. Ten minutes later the moon broke through a clearing in the clouds, and we spotted our destination. We'd already passed Tufutafoe, but could still see the shiny sands of the malae in the darkness behind us. Five more minutes and we would have been too far away to see it, so luck was changing in our favor.

Both boats crossed the reef over a deep spot just outside the village, my boat crossing ahead of the other one. We hit bottom twenty feet from shore, tied off the boats so they wouldn't float away when the tide came in, and went ashore, weapons at the ready. A half dozen men did a sweep of the perimeter to make sure there weren't any lookouts but found nothing. We were alone at the former village of Tufutafoe, hopefully a sign that the cannibals weren't expecting us.

At the center of the malae, Sina quietly gathered everyone for last-minute instructions. They were given in low tones, not much more than whispers, and from this point on, there would be no more talking. The rain that had been such a problem before was now working to our benefit. A wet jungle is softer and quieter, so it would be easier to move into position without awakening anyone. Or so we hoped. Our plan required that we surround them, make them think we had a huge force coming against them from every direction, and take as many of them alive as possible.

Plans, of course, are one thing. The reality is usually something else.

Sina was to lead half of our fighters along the foothills above where the enemy was camped. Once they got above them, half of her group would move on to block off escape to the south. I was to lead the other half through the jungle below their encampment, leaving half on the northern side to prevent escape there. Once everyone was in place, we would commence our attack, screaming our war cries at the top of our lungs. Hopefully, our lines would be so tight that none of the cannibals would be able to escape through them. The problem was in the timing. We had no way of knowing when everyone would be in place. All we could do was guess at the time it should take to get there, and hope for the best.

It had taken us over an hour of walking time from Tufutafoe to their camp, but that was in daylight and along sort of a trail. Trying to do it at night and in the jungle rather than on the trail, I thought, would likely take at least twice that long. The People didn't have watches or any concept of what an hour was, so Sina explained that the attack would commence when a star, which they all seemed familiar with, had moved to a certain point in the western sky.

Sina, despite being a warrior hardened in battle, showed a great deal of compassion when she assigned Apu and Little Sina to accompany me. That didn't guarantee their safety, of course, but she knew I'd feel better if I was protecting them rather than some villager with little battle experience. They shouldn't have been there in the first place, but when we were recruiting every other person healthy enough to hold a weapon, I couldn't leave them out with a clear conscience.

Sina and her group, which included Masana, took off up the hill, avoiding the trail so the chances of encountering an enemy were reduced. Me and my group dropped into the jungle several hundred yards below the trail and made our way quietly through the wet undergrowth. Although Sina's group was on the slopes of the foothills, the jungle was as thick there as it was on the slopes below, so she didn't have it any easier than we did. Which wasn't easy. Anyone who's traveled through jungle knows how difficult it can be. Doing it at night while it's wet made it a lot worse.

Not only was the vegetation wet, but the lava rock below it was wet as well, making it slippery and dangerous to walk on. Most of us fell more times than we cared to admit, occasionally getting gashes in our knees or ankles, but fortunately nothing bad enough to stop us. We had to be completely silent, even a gasp when we slipped could warn the enemy of our coming, and my group managed it beautifully. It was hard to tell when we reached the point that would put us directly beneath the enemy camp, so I sent half my crew to sneak up to the trail, hoping it would put them in the right place, to the north of the enemy camp. They had to be near enough to make their move when the time came, but not so close as to warn them of our presence.

The rest of us began the slow move up the slope, hoping to come directly below the enemy. It was slow going, and the star that was to mark the beginning of the attack was quickly getting in place. We almost had to crawl to keep from tripping and alerting the enemy. The wet vegetation soaked us to the bone, our hands were wet, muddy and slippery, but we kept going. Only small patches of moonlight penetrated the canopy of trees, and twice a man went up a tree to see the location of the star. We had to be there on time. Everyone had to be in place when the action started if it was going to work.

I was beginning to think we'd missed our assigned position. We were getting close to the foothills and still no sign of the enemy camp. Then it appeared, less than twenty feet in front of us. The rockslide, with men sleeping in recesses, on flat rocks, sitting against boulders, or sprawled out in the limited open areas. There was no movement. Nothing indicated that they had guards on duty, or that they were aware of us. It was deathly silent except for the snoring sounds that carried loud and clear in the air. I hoped beyond hope that the others were in their assigned positions. In order for the plan to work it had to be a coordinated attack.

Apu was terrified, wiping his hands constantly and fondling his spear in jerking motions. It wasn't his first fight, but by far the most threatening. Little Sina was scared, but more composed than her father. The rest of the group fidgeted with their weapons, but waited as the moments slowly tickled by, ready to do what had to be done. I'm always afraid when waiting for a fight, but once it gets started I seem to act on instinct, with no attempt at thinking things through. I respond as the situation dictates, and so far I'd been lucky. Would I be lucky again this time? Could I protect Apu and Little Sina against armed combatants who were more experienced in warfare than they?

Suddenly, without warning, a series of shrill war cries rent the air above the enemy camp, followed quickly by another from my group to the north. We joined in below the encampment, but I didn't hear anything from Sina's group to the south. As a united front we charged into the encampment, wielding our spears and ready for a fight. Some of the enemy were scrambling to their feet, grabbing up nearby weapons and searching the darkness for the attack that was upon them. When we entered the rocky clearing they could see us, but in the darkness we were just threatening shadows in the dark.

What they could see appeared to be a big mass of armed enemies coming for them. War cries sounded from every direction, letting them know they were surrounded.

At first they panicked, not sure where the bulk of the threat was coming from. Those who'd taken up weapons tried to stand their ground, but weren't sure where that ground was. The stupid ones who'd neglected to take up weapons, not hearing any threat from the south, started running in that direction. Just then our warriors there exploded from the jungle, finally adding their war cries to the shrill noise that seemed to pierce the sky. Cut off from the south, those who had sought escape in that direction were turned back. Too late, some of them tried to grab up weapons, but didn't know which direction to point them.

Sina's booming voice filled the night. "Put down your weapons and live, or fight and die" she screamed. The enemy warriors froze at the sound of that demanding tone, and for a moment it seemed like they were going to obey her, until the one I assumed to be their leader raised his club and charged. The rest followed suit.

They weren't certain which way to charge, though, so their confusion became our ally. So far, things were going according to plan. It looked obvious that we were going to have to kill some of them, but I had no doubt they'd surrender once they realized how vulnerable they were. Unfortunately, the leader turned west, toward me and my fighters, followed by two others. The rest turned wherever they saw the nearest warrior, which was everywhere. Those I would have to leave for the others to take care of. I had enough to deal with on my side of the fray.

The leader came at us in the dark, confused and able to see only the shadowy forms of his enemy. His angle put him off to my left about twenty feet, so I started after him, trying to run over uneven lava. It wasn't easy. Then a high-pitched war cry sounded as a young warrior confronted him. My blood froze, realizing that Little Sina was there, taking on a skilled fighter, and I couldn't get there in time to help.

As I was sprinting toward her in the darkness, I saw her spear arcing toward his head. It was blocked by a powerful blow from his club, parrying her attack. It knocked her spear hard to the side, and she barely managed to hang on to it, but she was driven off balance. As she tried to regain her battle stance, she tripped over something and fell.

My worst fears were realized as he lifted his club to finish her off while I was still ten feet away, screaming my war cry at the top of my lungs. He ignored me, lifted his club for what would certainly be a fatal blow, when another spear suddenly struck the back of his head. It wasn't a strong enough blow to kill him, but it dazed him and took his attention from Little Sina. I was close enough by then to see Apu standing to the side of the brute, defending his daughter at his own peril. The cannibal was just raising his club against Apu when another spear embedded itself in his stomach. This time it was Little Sina, still struggling to get back to her feet, defending her father.

The man was hurt but not down, and by then I was there, delivering a stroke to the side of his head. It was a solid blow, the teeth of my spear digging into his brain, and he fell without a sound. There wasn't time to examine him for two others were closing fast. I turned on them, found Apu on my left and Little Sina on my right as we engaged the oncoming fighters.

One of them charged and I crossed weapons with him, the other one stood back, uncertain. The club of the man I was fighting fell to the ground almost immediately when my spear ricocheted off his hand. He quickly knelt in a posture of submission, choosing to surrender rather than die. Apu rushed in on him, his spear raised and ready for a killing strike, but I called him off. We needed to take as many of them alive as possible. The other one, seeing that the other one would be spared, dropped his weapon and surrendered as well. Both weapons were grabbed and thrown way out into the bush, well beyond their reach.

Little Sina and Apu stood guard over those two while I went to assist anyone else in need. Sina was crossing her blade with a man wielding the typical club of the cannibals, but it didn't appear that she needed help, so I went on, searching for anyone in danger. One of our warriors was down with a nasty cut on his side, a cannibal standing over him ready to finish him off. I screamed my war cry at him and attacked, distracting him from his blow. He turned to face me, but too late. My spear hit him before he could swing, breaking bones that protruded through the flesh of his forearm. His weapon dropped from his crippled hand. He cried out in pain and turned to me in disbelief.

"Surrender or die!" I shrieked at him, my blade up and ready for another stroke that would kill him. He nodded, dropped to his knees, and draped his broken arm against his body. I threw his weapon also into the jungle. The man he had been about to finish off tried to stand, but he was in agony. It wasn't a fatal wound, but painful. With enough care, he would recover, but for the moment I couldn't tend to him. There was still a lot of fighting going on, and I was needed elsewhere.

On the southern end of the clearing, one of my warriors was down, the side of his head bashed in. There was no help for him except to take on the cannibal that had bested him. He didn't see me coming in the darkness, so I thrust the tip of my spear into his back. Not enough to kill, or even to incapacitate him, but enough to hurt and let him know I could have done a great deal more. He swung around, raising his club, but the wound in his back made the movement painfully difficult.

"Surrender or die!" I repeated, my spear up and ready. Two other black fighters nearby dropped to their knees in fear, thinking I'd made the demand of them. The one I'd injured didn't surrender. In his eyes, there was defiance. He wanted to fight, despite the pain that restricted his ability to do so. For a moment he just stood in the dark trying to make up his mind, and then he brought his club up and attacked with a shriek of pain.

His blow was more powerful than I'd expected from a wounded man, and I was barely able to parry it with my spear. I used the momentum of my stroke to come back at him on the other side, but he was fast, and managed to knock aside my swing. His club came down again, aimed at my head. I was able to parry it away from my head, but it struck a glancing blow on my right shoulder. It hurt like hell, but did no serious damage. He saw my grimace and assumed I was hurt more than I was. A grin of victory crossed his features as he brought his club up for another blow.

With his arms raised to strike, I swung at his exposed mid-section. My blade went six inches deep into his stomach, just below the rib cage. He bellowed, dropped his club, and clutched his wound with a look of surprise on his face. I could read his thoughts. *This can't be. A woman killing me? No, it can't be.*

But it was. For long moments he stood there, trying to hold his butchered body together as blood flowed down his stomach and thighs, then onto the ground. His knees trembled, getting weaker, but he still stood there gazing at me with a hatred that wouldn't let him die. Finally, he fell to his knees, too late to do it in submission.

"You had a choice, foolish man. We offered to let you live, but you chose to die. It is you who has done this to you, not me." And then he slowly dropped and finished bleeding to death on his belly.

By then the fight was pretty much over. There were a lot of cannibals dead or wounded on the rocks, along with two of ours who had died. Three more of ours were wounded, two seriously enough that their survival was in question. Still, it was a clear victory for us. Most of the black fighters were captured unhurt, exactly as we'd hoped. More had died than we wanted, and the wounded would be a problem, but we had enough to hopefully negotiate a peaceful solution with those yet to arrive.

Sina found me in the darkness, clearly unhappy as she surveyed the dead man in front of me. “You kill one of them?” she asked in frustration.

“Two,” I replied. “I’m sorry about that. They didn’t give me a choice.”

“You kill two, I kill none,” she sighed. “Bad plan. Very bad plan. If my spear drinks no blood it is a bad plan.”

I laughed with her. “Yes, I think it was a very bad plan. Next time the plan will be for you to kill two, and I will kill none. Would that be a good plan?”

“It would be a good plan,” she shot back, teasing. “A spear is not made for peace, but to drink blood in war.”

“A spear is made to protect the peace,” I countered. “Hopefully, that is what we have done.”

“We will see.”

“Yes, we will see,” I nodded.

Chapter 23

It was a long trip down the trail back to the boats. We had to guard the captives, some of whom were carrying their wounded companions. Some of our people were likewise carrying our wounded and dead, which were not left behind. The dead from among the enemy force had been left on the field of battle. Another black man died on the way and was left for the insects to devour in the jungle. Our dead were being returned to their families, and our wounded taken back to hopefully be healed.

In the aftermath of the battle, when the enemy was disarmed and under guard, they were dismayed at the force that had conquered them. Our numbers were fewer than theirs and consisted largely of women, girls, and boys not yet grown. Only a handful of grown men, suitable to be warriors, made up our force. And, of course, two women who were alleged to be warrior goddesses in the legends.

My shoulder ached from the hit I'd taken, but I tried to ignore it as we moved down the trail toward Tufutafoe. Once we got there, we had to pull the boats in closer to shore because the tide was in and the water was five feet deeper. The captives were loaded, half on each boat, and then we headed out to sea and back toward Falelima. One of the captives tried to escape before we reached the reef and had to be tied up to avoid further attempts. The rest accepted their fate, not knowing what we had planned for them. They probably thought we intended to eat them, and we didn't bother to tell them otherwise.

There were two more boats at Falelima than there were when we'd left, which meant one of the villages had returned during the night. It was just getting light when we arrived, and everyone came out to greet us. The captives were taken to the malae and securely tied with strong cording, on their knees or in sitting positions near the shore where they could be seen by anyone approaching. Helpless, they had no choice but to await whatever fate was in store for them.

So far there was no sign of the approaching enemy boats, and those of us who had spent the night capturing the enemy were exhausted, so we got some rest while the recently arrived warriors stood guard over the captives. Not that they were going anywhere, being tied the way they were.

Sina was still upset over the fact that the enemies she confronted during the night had surrendered rather than putting up a fight, keeping her from making any kills. The warrior in her didn't understand a fighter who wouldn't fight, but she accepted it grudgingly because of the crucial role the captives would play when their arriving party got here.

We were served a meal after we woke up, and the leftovers were given to the captives. No doubt they thought we were just fattening them up for the pot, and we let them continue thinking that way. The lookouts on the cliffs above the village were keeping an eye out for the enemy boats, so until they announced a sighting, we were able to relax a bit. A dozen of the newly arrived warriors, while keeping a close vigil over the captives armed with spears, javelins, and clubs, taunted them with jests about what we planned to do with them. Without actually saying what it would be, they hinted at extreme torture and pain.

We had estimated that the enemy force would number about twenty, and thought that between eight and ten had been killed, but we ended up with twenty-nine captives. Their force had been much larger than expected, and our nighttime raid had been unexpectedly successful. We'd been lucky, and we all knew it. Had they been prepared for us, we would have easily been overwhelmed and defeated.

Some of the guards took pleasure out of keeping the prisoners in a state of fear. There was no mention of our true intentions, just hints of painful torture and slow death. Although it seemed cruel, we didn't interfere. The more fear we could instill in them the better when the time came for negotiations to commence with the incoming forces.

At high noon, there was still no sign of incoming boats, and the captives had grown tired of being on their knees. Most were lying on the sand, some sleeping while others simply awaited what lay ahead with abandon. For the most part, they were brave, but some were less courageous and openly wept in the sand, expecting the worst. By midday, there was a feeling of doom that saturated the enemy troops who were waiting to die in agony.

Then, late in the afternoon, the alarm was sounded. The boats were in sight.

An early count, made while they were little more than dots on the horizon, indicated fewer than thirty boats. It was a large force, larger than we'd expected, and there were likely more coming that we couldn't see yet. They were still far enough out that an accurate count of the ones that were visible wasn't easy to make, and it was likely there were others as yet impossible to see. It wasn't good news.

The captives, once they became aware that others of their kind were coming, possibly even their wives and children, began to take hope. They believed they would be rescued by the reinforcements, who would swarm over our meager forces. Rather than us eating them, they imagined they would eat us. They were back on their knees, watching their people draw ever closer with renewed hope and determination. Instead of us torturing them, they planned to exact vengeance on us as soon as they were liberated.

While we watched the slow approach of the enemy force, two boats from another of our villages arrived, which added fairly significant numbers to our side. Even so, the oncoming boats would bring more than enough numbers to overpower us. We had to assume there were skilled warriors on those boats in addition to the women and children we assumed would be with them, and could only hope that their desire to save our captives would be stronger than their wish for battle. It was a time of stress, watching those boats come nearer and nearer. The number of boats went up to over forty as those in the rear came into view.

To say we were afraid would be an understatement. We were watching what would likely be our demise as those boats came closer and closer until they were just beyond the reef. Then they stopped. From there, they could see everything. Our small village, the captives who were now gagged as well as tied, and our warriors who were guarding them. The rest of our forces were hidden from view and would remain so as long as it took to complete the plan, making them guess as to how many of us there actually were. The night came on, darkness fell over the malae, and still, the boats remained beyond the reef.

Torches were lit and placed around the captives. None were used in the village itself, or by those of our warriors who remained hidden among the trees and jungle surrounding the village. There were torches on the boats, however, making it possible for us to observe them without difficulty. Our luck held out. The sky was clear, the light of the moon and stars shining brightly. We could see them fairly well out on the open sea, but they couldn't see into the jungle's shadows where our forces were hidden. If we got the chance to actually engage in negotiations, we didn't want them to know how small our force was. They would have to guess at that, and the fact that we had so many captives would suggest a bigger number than we had in reality.

Throughout the night, the boats just sat out there beyond the reef, apparently waiting for daylight to do whatever it was they planned to do. Each boat contained thirty or forty people, but it was difficult to tell in the darkness if they were all warriors or family units. We could tell that there were a few children out there, but most of the ones we could see appeared to be adults. Women or men? It wasn't easy to tell at that distance despite the light from their torches. Their men wore hair as long as their women, and their black garb blended into the darkness, making it difficult to tell one sex from the other.

A long night of vigilant watching revealed no movement from the boats. Were they just waiting, hoping to increase our fear? If so, it was working. Our fighters were nervous, afraid of what might come when daylight flooded the field, and rightfully so. Their numbers were clearly greater than ours, probably ten or more to our one. In a battle, we would be defeated, and there was nothing we could do about it. Our only hope was that they would want our captives alive and willing to leave the island in order to keep them that way.

Morning broke clear, sunshine flooding the village and malae, where many of the captives were now back to their knees. They'd slept fitfully, scarcely more than short naps, expecting to be saved by the forces now arrayed beyond the reef. Our guards remained vigilant, with orders to slaughter the captives if an enemy force should try coming ashore. There would be plenty of enemy warriors to fight against without allowing the captives to be among them.

I didn't like the idea of killing helpless prisoners, but I didn't see any way around it. If a fight was forced on us, we couldn't let our captives join the already overwhelming forces arrayed against us. Our numbers were simply too small to allow the odds to be turned that much more in favor of the enemy.

It was necessary to keep the prisoners gagged so they couldn't reveal the true strength of our forces if negotiations actually got underway. Short pieces of tree limbs about an inch and a half in diameter were put in their mouths crosswise, and tied in place with cording. They could make screaming sounds but couldn't form words. Those who were wounded could sob and reveal their agony, which might help as well.

Other than those guarding the prisoners, everyone else kept out of sight. The village looked empty, forcing those on the boats to guess at our strength. It was a small village, which might make them think we were few, but there were six huge boats lined up along the shore, each capable of holding thirty or more warriors. That would give them plenty to think about, and keep them wondering just how strong our forces were. Maybe that was why they'd spent the night beyond the reef. Were they watching to see how many of our warriors they could see? If so, they hadn't seen any more than those guarding the captives.

The mere fact that we had captured so many of their warriors alive suggested a force much larger than the guards watching over them now. Maybe they thought there were others still alive from the advance team of over a hundred. If so, where were they? Why weren't they coming to the rescue of our captives? Were they hiding in the jungle, afraid to face us? They had to be envisioning all kinds of situations, and all of them seemed to lead to the conclusion that we were waiting for them with a strong enough force to not fear them.

The women and children who were unable to fight had been taken to their retreat, guarded by two women with spears. Little Sina had refused to take on that duty this time. She was fueled by anger after being dropped to the ground and nearly killed in the nighttime raid. That, she swore, was not going to happen again. She was brave, but it was that very aspect of her that scared me. Being foolhardy would lead her into a situation where her skills were insufficient to save her, and next time there might not be anyone there to help.

I knew what it was like to go into battle as an unskilled warrior. In my case, it had been luck that helped me survive on that first trip back into Sina's time. And it had been luck that spared Little Sina during the night, when Apu was there just in time. But luck can only be stretched so far. It's an unreliable companion, an ally that may not come at the critical moment. This fight, if it came, would be too intense to hope for luck. If war broke out, it would be a miracle if any of us survived. And any who did survive would probably wind up in an enemy cook pot anyway.

Again, I wondered what the implications of time travel would be if we were to die in a time to which we weren't born and didn't belong. That would be a direct change to the future that was our rightful place, wouldn't it? Which meant that the events of history were subject to change, and not only in small ways. If we lost this battle, didn't that mean the blacks would occupy these islands rather than the Polynesians? That would be a major shift in the history of the world. Polynesia, if it existed at all, would be located on the islands far to the south, closer to what would become New Zealand or Australia, where the Melanesians and Micronesians were located in my time.

My reflections on the impact of historic change were interrupted by a single boat approaching from the armada beyond the reef. It contained only a dozen or so blacks, suggesting that their intention wasn't to commence a war, but to talk. So far so good, as long as the talks went the way we wanted them to. We were getting a chance to arrive at a peaceful solution. Now it was up to us to make it happen.

The boat, pushed forward by the waves, stopped when it hit bottom about ten feet from the shore. A single man stood as if in ceremonial fashion and waited. I stepped from cover and approached cautiously, wary of a trick. He was arrayed in some kind of headdress, resplendent with feathers and colorful tapa cloth. His robes were bright red, covering the entire length of his body. Around his neck was a whip-like device, made of dried coconut husks, much like those the chiefs of The People wore. His feet were bare. He was, without any doubt, one of their highest chieftains, which also worked in our favor. We would commence negotiations with a man capable of making decisions.

There was no formal greeting when I reached the shore. He just stood there in silence for a moment, looking me over with clear disdain. I was a woman at a time when men dominated such things as warfare, so he wasn't going to take me seriously. Not yet, anyway. With luck, that would change in the next few moments. It might have been better to have a man undertake this function, but none of the chiefs in the village were willing or, according to both me and Sina, able to do it.

I did likewise with him, saying nothing for a long moment to let him know I was examining him just as he was me. It was a long, tense moment, both of us trying to decide the best course of action. I wanted him to speak first, to be the one who submitted himself to ending the silence. Finally, he did.

“Why do you have our men tied and gagged?” he demanded.

“The village of Falelima extends warm greetings if you come in peace. If you come for war, we will greet you in like manner. I would address you with the proper respect, but I know not your name or title,” I replied. “Am I allowed to know these things?”

"Do you not see we are many?" he shot back.

"It is of little concern," I said, keeping my voice soft and friendly. "If you come for war, we will greet you in like manner. That is my word, and it is a true word. You do not see how many we are, but this you can see. You sent many against us, and most of them are dead. These are all that remain of those who came for war. Do you wish them alive, or should we kill them also?"

"Why would you kill those who are bound?"

"Why would I let them live to fight against me? I have already fought and defeated them many times. It would be useless to do so again."

For a minute he was silent, obviously deep in thought. It was clear that he wanted them alive, which worked in our favor, but would it be enough? Would he abandon his desire to take these islands in order to spare them? That was yet to be seen. Finally, he spoke.

"One of your captives is my son. I would have him alive, and swear that he will not again raise his weapon against you."

"He has no weapon, chief, if that is what you are. None of these men have weapons, and we will not allow you to give them such. Which of these men is your son?"

He pointed out one of the youngest, sitting cross-legged, his arms bound behind him, his ankles bound as well. He had a gash on his forehead, likely a wound he got in battle.

"He has suffered a wound, it seems. I will tell you a thing about this man. He will be the first to die if you do not accept our terms."

"What are the terms of which you speak?"

"You and all those with you will return to the lands from whence you came and never return to our island again. You will not raise weapons against us, and will leave in peace. Give me your word to do this, and we will surrender these men to you."

"Your terms are strong, woman. It is a long distance we have come to make villages in which we can live. It is not an easy thing to turn back after such a long journey."

"We, too, made a long journey to build our villages here, and it is not an easy thing to give them up now. These are our islands, and you will find no rest here."

"Perhaps we could go to another part of the island?"

"You are cannibals. You eat people. There can be no safety for your people or mine if we live on the same island. We have taken all the islands in this part of the ocean, so you must return to the islands from which you came, or watch your men die, and die yourselves if you make war with us, for we fear not your numbers."

"Do you think you could defeat us? We are strong in the art of war."

"These men," I said, pointing at the captives, "thought they were strong, but now they are bound and kneeling before us. You will find us very strong, as they did, if you come against us. If a war is your wish, let us begin it now, but remove first your women and children. We have no desire to kill them unless they take up arms against us. Anyone with a weapon will die. This is my word, and it is a true word."

"Your words are brave, but we fear no force led by a woman."

"We are led by two women," a voice I knew well sounded from behind me. Sina emerged from her hiding place to stand at my side, her spear clutched in her hand next to mine. "I am Sina, war chief of Tufutafoe, war goddess of legend and defender of these forests. And this," she indicated me, "is Salini, warrior goddess of legend from the village of Siata and adopted daughter of Tufutafoe. For thousands of years, we have never been defeated in battle, and will not be so now."

"We will not be frightened by two women, or a thousand. We are men trained and tested in the art of war. Women die easily when confronted by a man. This is my word, and it is also a true word."

Sina lifted her left arm straight into the air, grinning. The men guarding the prisoners raised their weapons menacingly over the heads of the captives. "And this is my word," she spat at him. "When I lower this arm your men on the malae will die, and you can bring your warriors ashore to die with them. This is a more true word than yours."

There was another long pause. The man on the boat had a lot to think about, it seemed. He tried to cover his discomfort, or fear, but it floated on the air like a vapor and couldn't be hidden. I didn't know if he feared for all the prisoners, or just his son, but I took that opportunity to push him yet farther. I turned to the guards and told them to bring the chief's son to his feet and put a knife to his throat.

"I think," I said, "that this man, who you say is your son, will be the first to die, for his father is an arrogant, boastful man who wishes to lead his people to destruction. No son should live with the disgrace of such a man as his father. We will spare him the dishonor by killing him now."

"If you do this we will make war with you."

"It is your intention to make war with us anyway," I pointed out, "so it is of little consequence. On my word, we will spare your son the humiliation his father will bring to his people by cutting his throat. This, also, is my word."

"You want us to leave these islands and go back to our lands to the south? It is a long journey, and we have little food and water left. Such a journey would not be possible."

"If it is food and water you need, we will provide it for you, for we have no desire to kill you, but you will not remain on any island near here and live. If you return to your own lands, you may take our captives with you, but if you try to remain, they will die first, and then you and your people will do likewise. So tell us now whether you and your people will live or die."

This time the silence lasted much longer. There was no response for at least five minutes, but neither Sina nor I broke that silence. We wanted him to do that, to show his intentions, his fears, or his boldness. The next person to speak would likely be the one to cave in, so we waited, trying to act calm and brave.

And waited.

Chapter 24

His response, when it finally came, was given in a subdued tone. There was no bravado in it, no blustery demands.

"I must confer with my chiefs."

"You are free to do so," I replied, still trying to sound commanding. "But before you go, tell us your name and title so that we may address you properly upon your return."

"I am Po'owahe," he said, "high chief and king over all my people."

"Then go in peace, King Po'owahe, and return in peace with the decision of your chiefs. Let us not start a war that will mean the end of your reign if it can be avoided. We await your return, but you must return as you came, without a force at your side."

Those who were manning the paddles on his boat started the trip back to the fleet beyond the reef while Sina and I stood straight but quiet on the shore. We wanted him to think we didn't fear him or his warriors, though the opposite was the reality of our situation. We were terrified of what would happen if he brought his forces against us in a battle we couldn't win.

It took the better part of an hour for his boat to reach the armada in the deeper ocean, where the swells were making it difficult for them to maintain their position. By then it was approaching midday. The captives, tired of kneeling, were laying in the sand again, resting their tired legs. Po'owahe's son was agitated, trying hard to speak despite the limb that gagged him. I told his guard to remove the gag so he could speak, which he did without delay.

“My father will not surrender to you, woman. He will kill you and feast on your flesh, you and all your people. He is a warrior of renown among our people.”

“If that be the case,” I told him, “you will not live to see it.” I motioned for the guards to gag him again and put him down on the malae with the others.

A short time later two more boats of our villagers pulled in, coming along the coast inside the reef to avoid the enemy boats farther out. That added nearly twenty more warriors to our ranks, and the enemy must have surely watched with dismay as more warriors joined us. It was a good sign, but we would still be greatly outnumbered if war was declared. As soon as the new arrivals came on shore we told them to conceal themselves in the jungle along with the rest of our warriors. The enemy would have to guess as to how many of us were out of sight.

We ate our lunch and fed the prisoners, gave them water and coconut juice, then settled down to await the return of their king, hoping for the best. Po’owahe, we felt, was crushed, and would concede to our demands. But the other chiefs, if their system was similar to those of The People, might change his mind or force him into a war he didn’t want. Everything hinged on his meeting with the lower chiefs, and we had no way of knowing how that might turn out.

He returned sooner than expected. With him were six men on the paddles and two others, probably a couple of his chiefs, who stood straight and proud as the boat came to shore. I went out to greet them as I had Po’owahe, with Sina at my side.

“We greet you, King Po’owahe,” I called out. “It is a greeting of peace, if such is your decision. If not, it is a greeting of enemies who will soon crush you.”

"Hold, woman," he called back as the boat hit bottom and stopped over twenty feet out due to the receding tide. "I have an offer for you."

"Speak, then, King Po'owahe."

His demeanor turned haughty again. "You have said you will crush us, but you are women. Can a woman defeat a true warrior, a man experienced in warfare, who knows the art of battle?"

"If you come against us you will find out, Po'owahe," Sina shot back at him.

"Here is my offer, then," he went on. "There are two women before me who claim to be warriors of renown. I will send two of my men to do battle with the two of you. If we are victorious we will commence a war with your people. If they die in battle we will leave these waters and return to the lands we have left behind. Is it an acceptable offer?"

"Why should I let a single warrior from your people land on this island?" I retorted.

"If you are the goddesses you say, you should fear no such threat," he replied, "for a goddess cannot be bested in a fight."

"It is not fear I feel, King Po'owahe. I have shed much blood and grow tired of it. I have slain the Aun'u, who were cannibals like yourselves. I have slain dangerous beasts and gods of great strength. Why should I want to shed the blood of your people?"

"That is my offer," he said, standing erect in the boat. "If you refuse it, we will see your fear and make war with you."

It was Sina who spoke next, saying exactly what I would have expected from her, but didn't want to hear.

"Bring your warriors, king of the dead, and when they die so also will those who are bound on our malae. And if you bring war upon us, you and your people will die as well, for we have so spoken."

It wasn't a good situation for Po'owahe. I could see the hesitation in his eyes, a desire to protect his son, and perhaps the rest of our captives, but he was trapped. He had been forced to make the offer, and with Sina's acceptance, he was compelled to carry it out. I wanted none of it. Maybe I could defeat the warrior they would send against me, maybe I couldn't. If I failed, The People would be in grave danger. It wasn't any better for me and Sina than it was for Po'owahe. Too much hung in the balance for both sides.

"They will be given safety while they come ashore?" he asked, probably hoping for a negative response, which would alter the terms of the challenge.

"They will reach the malae in safety," Sina blurted, "but when they raise their weapons they will surely die."

I wished I could be as optimistic as Sina, but there was within me a concern for what the contest might bring. The two warriors who stepped from the boat into the shallows were armed with clubs. They were heavy weapons not easily parried with a spear and could do severe damage. An injured arm or leg could spell the end of the contest. A blow to the stomach or shoulders, or any other body part, would be disastrous. This was a battle we had to win. Too many lives were at stake, not just our own.

My mouth was dry as the two black warriors stepped from the water, their clubs clutched tight and ready for a fight. My palms were sweaty, my forehead dripping salty water into my eyes. Usually, a fight was brought on me suddenly, without time to think. This time it was a slow advance, allowing too much time to ponder how it might go. The two men were tall and muscular and wore nothing but long breech clouts. No shirts, no shoes or sandals. Tattoos covered their chests, stomachs, and thighs, making them look rather fearsome. When they were before us I spoke to the warriors.

"I would know your names and titles," I demanded, "so that I will know who it is that I have killed. You should not die without a name."

"I am Wahano," the first one said, "and this is Pohiwo. It is said that you are goddesses of war. Is it true?"

"You will know shortly," Sina broke in, "though you will die in the knowing."

"Then we will ask your names, that we will know what goddesses we have killed."

"I am Sina, war chief of Tufutafoe, legendary goddess of war, and defender of the forest. And this is Salini, warrior goddess of Siata, adopted daughter of Tufutafoe, who has fought with me through thousands of years, and will do so again long after you are dead and gone."

"We will see," Wahano said with a chuckle. "I see no goddesses, only women. We will kill you, and then we will kill your people. That is what I see."

"Then your eyes are weak, warrior," Sina taunted. "Raise your silly club and prepare to die."

In an instant, he did exactly that, bringing his club up and rushing at Sina. Pohiwo, following suit, raised his club and came at me, preparing to deliver a crushing blow to my head. From that second on, I was focused totally on Pohiwo and lost track of the battle between Sina and Wahano.

I was able to parry the first blow, knocking the club aside, but we both almost lost our balance in the soft sand. He recovered quickly, making a wide swing with that deadly weapon at my midsection. I knocked it away just in time, and again he was off balance for a split second. Taking advantage of that, I brought my spear into an offensive plunge at his belly.

The club is, of necessity, a heavy weapon. It can do massive damage when it strikes, but is too heavy for quick defensive moves. His came sideways to parry my thrust, but a bit late. I didn't get him in the stomach as I'd wanted, but my spear left a nasty gash in his side. It was a painful wound but hardly one to put him down. His eyes grew wide, and his nostrils flared in anger over the fact that I was the first to draw blood. Hatred seized him. There was no way he could be defeated by a woman.

Wary now, he backed off a couple of steps to size up the situation and decide how to proceed. There was clearly no quit in him, only a greater determination to win. Blood ran down his thigh from the wound, but not enough to weaken him. If anything, he would be an even greater threat for having been injured by what he considered nothing more than a weak woman. But he had been injured by that woman, and would either become more cautious or attack with an insane frenzy.

"If you surrender you will be spared," I told him. "There is no need for a brave man such as yourself to die in a foreign land."

"It is you who will die," he shrieked, swinging his club in a sideways attack aimed at my hip. I was barely able to parry his blow, which was backed by strong arms, but managed it. Again, he was off balance for a split second, so I swung my weapon at his throat, trying to crush it with the teeth, but he managed to back away in time, and my spear sliced through nothing but air. While I was bringing my weapon back into position, he swung his club again. This time he didn't try to hit me with it, but my spear. The force of the impact knocked it from my hands and sent it flying.

It fell to the sand several feet away, and I stood before the enemy weaponless. That gave him a false sense of security, thinking the victory was clearly his. He raised his club again, another blow that was intended to crush my head. I jerked aside to dodge it, but not quite enough. That lethal chunk of wood was going to hit my shoulder, so I grabbed it with my hands as it came down. It would most certainly have been a debilitating injury, likely one that would have left me helpless. But it didn't turn out that way. I was able to stop it just as it touched my shoulder. That meant I had the heaviest part of his weapon in my hands, which gave me the advantage of leverage. Using that leverage, I shoved the club with all my might, smashing the handle of his own weapon right into his face.

Again, it was no killing wound, but it stunned him long enough for me to grab up my own weapon from the sand and face him again. Once more, I offered to spare his life if he surrendered, but I doubt he even heard my words. He raised the club over his head where blood was now flowing from his shattered teeth and nose, but before he could bring it into action, I once again attacked his stomach.

My aim was high, striking him in the ribs where my blade couldn't penetrate. He bellowed in either anguish or anger—I couldn't tell which—and brought his club down again. I was able to sidestep it, and before he could pull it from the sand, I stepped on it. My weight was too much for him to pull it free, but as I tried to swing my spear again, he jerked on it fiercely. I was thrown backward, lost my balance in the sand, and fell hard to my back.

I saw a look of triumph streak over his features. His opponent was down and helpless, the victory was his. He raised his club for the final blow, but in doing so, left his mid-section unprotected. I thrust my blade, still lying on my back, deep into his stomach. The shock and surprise made him jump back instinctively, but he managed to maintain his grip on the club, which was still held high. For a few seconds, he hesitated, then recovered and brought his weapon down hard at my head. I rolled away, raised my left leg to kick his arm. The club fell from his grip, dropping to the sand next to me.

He was now bleeding from two minor wounds on the head and side, and a more serious one to his stomach, but there was still no give in him. The wound to his stomach would eventually bleed him out, and it would be a slow and painful death. Still, there was still no quit in him. He lunged for the club, and while he was bent to retrieve it, I kicked his back as hard as I could from my prone position. He was thrown off balance and fell to the sand. I jumped up, brought my spear into position over his heart, and once again spoke to him.

"You have lost, Pohiwo," I screamed. "You will die in minutes from the loss of blood, so there is no need for me to shorten your life now. Surrender, or I will end you now."

It wasn't much of an offer, and I knew it. So did he. Yes, he knew he was dying, but he had to kill me first. That was all the crazed man could focus his mind on. Before he died, he had to put me down, and there was still enough life in him to continue the fight. I allowed him to regain his feet, his lower body and legs now covered with blood. He was slower, weaker, but still determined.

Without bothering to raise the club again, he swung it from the sand, aiming for my right shoulder. If he could disable that, he would certainly be able to kill me because I'd be defenseless, but I brought the butt of my spear up just in time. His attack wasn't as strong as before, and when the club was stopped in mid-stroke, it fell from his hands once again, landing at his feet.

He just stood there, wobbling slightly as he looked at me in disbelief. The battle was over. That realization entered his brain, and although he wasn't ready to surrender, he knew he was about to die and just stood there, looking at the woman who had conquered him in fair battle.

"Strike!" he demanded, wanting to die quickly rather than slowly bleeding to death. I turned my back to him, then whirled around with my spear up and swinging at his neck. The teeth dug in deep, severing his spinal cord. He went down without a sound, his death instantaneous.

There was no sound except for the sound of combat as Sina struggled against Wahano. I turned my attention there, no longer worrying about Pohiwo. He was dead, but Wahano was not. It wouldn't be right for me to assist Sina because of the way it would look, both to her and Po'owahe, but I desperately wanted to. I was winded from my own fight, but capable of more if necessary.

Wahano had a wound on his back, but it wasn't serious. I was disconcerted to see that Sina seemed exhausted, stumbling slightly in the sand as she shuffled her feet. Wahano was also tired, but still strong. Several times he brought his club down where Sina should have been, but each time she ducked beyond his reach. The ferocity of Wahano's attack kept her on the defensive, unable to strike back. Then, to my horror, she slipped in the sand and fell to her knees.

Wahano tasted victory and raised his club for the killing blow, but in that instant, Sina's spear came up. The blade went through his body just under the ribs, digging deep enough to cut vital organs. He jumped back, freeing himself from the weapon that had impaled him, trembled for a moment, wobbled on his feet, then fell to his knees, staring at her in disbelief.

"You are a strong warrior, Wahano," she told him, "but I have never been defeated. I am not happy to see such a brave warrior die, but you were offered life and rejected it." With that, she swung her spear at the side of his head, the teeth of which are so deadly, crushing his skull at the temple.

It was over. The warriors Po'owahe had sent against us were dead. Now we had to hope the king would honor his word to leave if we were victorious. I was tired and winded, but elated to be alive. With my spear in hand, I walked to the edge of the ocean closest to where Po'owahe's boat stood.

"It is done, King Po'owahe," I called to him. "Your challenge was accepted, and your warriors defeated. Now it is time to kill the captives, for that was the word of Sina. And after you watch them die, you will leave these islands, for that was your word."

He was speechless, unable to believe that his strongest warriors had been killed by women. I was somewhat surprised myself, but grateful it had turned out that way. His warriors had been strong and skilled, but the club is a less effective weapon than a good spear. It is too cumbersome to make a good defensive weapon or quick assault, designed mostly for offense by its sheer weight. In a war, however, it is wise to defend yourself as well as strike offensive blows. It had been our ability to defend ourselves that had won the day.

Po'owahe sat on the edge of his boat, defeated by his own challenge. "I will go and talk to my chiefs," he said, obviously not happy. "Do not yet kill your captives."

"There is no need for you to talk to your chiefs, for you are their king and made a bargain. If you come back with another challenge or say other than what you have already said, you will be slain over the body of your son. You may confer with them concerning the nature of your return to your lands, but no other thing must be discussed. If you speak with them of war, all of you will die. Even your women and children, for we will not care for them when they have no man to do it for them. That is my word, and it is a true word. It would be wise for you to keep that in mind when you talk to your chiefs."

"I will talk of our return to our former lands if you agree to the release of your captives. You promised to release them, did you not? That was your word."

"That promise was made when we demanded peace with your people. Instead, you sent warriors to do battle with us."

“Ayeee,” he shrieked. “It was a wrong thing I did. Take my life for it, but spare the life of my son.”

“We will spare him and the others. And we will spare your life as well,” I nodded, “provided you keep the promise you made. You may go, then. And when you return, leave enough space on your boat for the captives. Your other boats will leave before you, and after two days, when they are far away, we will release the captives to you.”

“Am I to be a captive for two days?”

“You are a king and will be treated as such while we ensure that your boats are well on their way. You will be treated with courtesy but will face restrictions as to where you can go. You may spend time with your son, though he will remain gagged like the others until the time they are released to you. A fale will be appointed to you, but you must not go beyond that fale and the malae. These are the terms. Do you accept them?”

“It is by my own words that I have no choice, and so I take my leave. I will return as I have spoken,” he mumbled, and told the men on the paddles to take him back across the reef, where the rest of his people were waiting.

Chapter 25

We thought we had the victory in our hands, but Po'owahe put a wrench in the gears when he didn't return with the boat that came back. It was a lower chief who returned instead, a huge man by the name of Titiama. He explained that Po'owahe was needed to organize the departure of the other boats and offered to stay in the king's stead. I turned to Sina, who was as unhappy about it as I was. We both had the same thought, and she acted on it immediately.

Going to where the captives were bound, she grabbed the son of Po'owahe by the hair and lifted him to his feet, facing Titiama. Then she drew her knife and placed it over the young man's throat.

"If Po'owahe does not come, I kill his boy. If he does not come, I kill all captives. This is my word. If he does not come, I kill all of his people on this island. You too, Titiama. Tell your men to go get the king or all die. You must stay until Po'owahe comes as he said. You and these," she said, indicating the captives behind her, "will die if the king does not come. This is my word, and it is a true word."

The black leaders were having a lot of speechless moments in the last day or so. This time it was Titiama's turn. He just stood there with his mouth agape, not sure what to say or do. This was more than he'd bargained for, more than he had the authority to deal with. If he refused to send for the king, he would be among those who died, which didn't really sit well with him. It was also pretty clear he didn't relish the idea of sending his boat back after failing in his mission. The consequences for that might be just as severe as staying.

He took one last look at Sina with her knife at the boy's throat, and made up his mind. "I go tell Po'owahe," he announced softly, almost timidly. "He will decide."

"Your men will go without you," Sina insisted. "If Po'owahe does not come, you will die. I will wait," Sina added, "until the shadow of the mountain comes to here. If Po'owahe has not come, I kill his son, and you, and all captives. Your men will tell Po'owahe of my word."

"They will tell him," Titiama nodded, instructing the men on the paddles as to what they had to do, and to go without him. We watched them go, knowing that it would probably take an hour or more before they came back. That would put it just a few hours before dark, and the one thing we didn't want was to have that fleet of boats just outside the reef during the night. We had made plans in case of a nighttime attack, but the fact remained that we were way outnumbered, regardless of any plan we put into effect.

Except for the guards standing over the captives, all our warriors were still hidden in the bush. They would not be showing themselves unless open hostilities broke out, even if Po'owahe returned as promised and the enemy boats took their leave of our shores. Sleeping mats, food, and water had been provided to sustain them throughout the vigil. They would stand by in case they were needed, but we couldn't risk letting Po'owahe see how meager our forces really were. There would be no reason for him not to turn his boats back once he had the captives and rejoined his fleet.

Shortly after Titiama's boat reached the others, we heard a tortured scream coming from his direction that was suddenly cut off. It seemed Po'owahe didn't like the outcome of his mission any more than we did, and took quick action against one of the boatmen to show his disapproval. Fifteen minutes later, the boat turned back again, this time with Po'owahe sitting in the front. He stood once they crossed the reef into calmer waters and remained that way until the boat touched bottom just offshore. He looked to make sure his son was alive before saying anything, then barked rather than spoke.

"You threaten my son," he bellowed. "Why do you do this?"

“We didn’t threaten him, King Po’owahe, you did,” Sina huffed. “You gave words of promise, but then you did not keep those words. Is this the way a king acts? We also gave words of promise, that if you did not keep your words we would kill your son and all captives, yet we did not do this, choosing instead to give you one more chance. Is it now that you will honor your words?”

He huffed, stomped his feet like a child in a temper tantrum, and raised his voice. “How do I know you won't kill me? Huh? How do I know this thing?”

“I have given my word,” Sina broke in, “and it is a true word. I do not go against my words, for I am not like you.”

“You say you are going to kill my son and captives if I don’t come back, but you did not do this. You dishonor your true word, yes?”

It was Sina’s turn to be silenced, but only for a few seconds. “What you say is true,” she admitted. “Now I honor my word and kill the captives, so you know my word is true.” She turned immediately toward the prisoners, her spear clutched tight and ready for action. Before she got five paces, though, the king called out in panic.

“No!” he shouted in frustration. “Your word is true, so it is now that I must make my words true. I will come, and you will do with me as you wish.”

“You will be treated well,” I reminded him, “for that is the word we have given. But you will go only where we allow you to go, for that is also our word. If you do not agree to this, you and your men will surely die. Do you accept this word?”

He nodded unhappily. There was a look of rebellion on his features, but he had no choice but to resign to our demand.

“Then you may come ashore,” I told him. “The boat on which you came must remain to take you back when the time comes. Your men must stay on the boat and cannot come ashore. Titiama will stay with them also. When your boats have gone beyond the sight of our lookouts for two days, you and your men will be free to leave. If you do not honor your word, you will die with your son and his companions."

It wasn't a harsh demand for his men to remain on their boat, for it was equipped for long journeys, with sleeping accommodations and food preparation gear. They would be no less comfortable there than they had been during their long trip across the ocean. The king slipped off the boat into the shallow water and waded to shore, where Sina confronted him.

"We welcome you, King Po'owahe, to our homes. It will soon be time for the evening meal, but before that, we will share kava with you in honor of your visit. There will be no fiafiaga, for many of my people want to see you dead, and I might not be able to protect you from them. I will stay with you during your visit to protect you from those who would do you harm. Your men will remain our captives, but you are a guest."

"I thank you for this kindness, the warrior woman who bested my most skilled fighter. Yes, you are indeed a goddess."

"I defeated only one," Sina reminded him. "Salini defeated the other. If you had not sent them against us, they would still be alive to return to their homeland with you. It was you who killed them, not us."

"I did not think a woman could be so strong in war," he said with a look of disbelief.

"They were strong warriors," she retorted, "for mere men. I give honor to them, but it was foolish to send them against us. You dishonored your word by doing so. If you are to be the king of your people, you should act as a king, with honor and words that are true."

"You have taught me a good lesson, warrior woman, and I will take it into my soul."

"This is good. Now go, King Po'owahe, and greet your son before we share the kava. He will remain gagged and will not be able to speak to you, but you may speak to him."

Neither of us escorted him to his son. He was free to go that far, and on to the fale beyond, which had been set aside for him. Those were his boundaries. He couldn't go beyond the fale or into the bush on either side of the village, where our warriors were hidden. Those who were guarding the captives didn't prevent him from going to his son but stood back in case of trouble.

We did not allow the king to remove the gag on his son or any of the others for fear they would reveal the true strength of our forces. He would find out, of course, when the captives were loaded onto his boat and sent out across the ocean. We hoped by then it would be too much trouble to turn back and come at us again. His people in the other boats would be at least two days ahead of him, so it would take a week or more for him to catch up to them. Would he bring them back against us then? We had to rely on his word and prepare for the worst.

He showed no compassion to his son. It appeared to him that being captured alive was nearly as bad as being killed, so his greeting was anything but cordial. He laid into him for being a poor warrior, for allowing himself and the others to be taken alive, which meant they hadn't put up enough of a fight. He didn't know how things had gone down, and didn't care. They had been captured rather than being killed. That was enough for him.

After a few minutes, I'd had enough of his jabbering on about fighting to the death, about the humiliation of being taken alive, so I went over to intercede.

"You know nothing of the bravery of these men, or how they came to be captives, Po'owahe. Your son was brave in battle, but we had many against these few, and to fight on after so many had died was futile. Perhaps you should remember that you, too, have been made to surrender yourself to our demands. You are no better than these," I said, gesturing to the captives, "so do not fault others for what you yourself have done."

"You said I was your guest, not your prisoner."

"This is what you are, because we are merciful. But you are our guest not of your own wish, but because we demanded it to be so. You, too, were forced to do as you have done rather than being killed, as we swore to do if you didn't honor your word."

"You are a hard woman."

"It would be well for you to remember that, Po'owahe. I am a warrior not accustomed to making my enemy a guest in my village. Do nothing beyond what we allow, and you will fare well. Otherwise, you and your men will surely die. This is a true word, so heed it well."

That quieted him down. He sat next to his son in the sand, saying nothing while I went to see about the preparations for the kava. It was well underway by then, taken care of by Sina. The rest of the villagers were still in hiding, and with the exception of a few women needed to prepare the evening meal, were not to be seen.

It was the simplest kava ceremony I ever saw. It consisted of me, Sina, and King Po'owahe, sitting in the sands of the malae. There was no high chief to preside, no taupou to serve, no pole of honor, and no speeches. Sina half-filled a polished coconut shell, the traditional cup for the kava ceremony, and offered it to Po'owahe. He drank it without spilling some as an offering to the gods. I doubt that he understood that custom among The People, so he wasn't chastised for it, although Sina was somewhat offended.

Sina gave me the cup next, and I dripped some over the front of the cup with the traditional words, then drank. Kava is an acquired taste, but one I had come to enjoy. The flavor is something like liquid metal in dishwater, but it's quite refreshing when you get used to it. The effects are also nice. It numbs the tongue first, then the mouth, the throat, and then the whole body. It isn't intoxicating, doesn't slow down the responses of the senses or physical abilities but makes things seem much more relaxed. Indeed, kava is nothing more than a muscle relaxer.

Sina drank last, again making an offering to the gods, then passed the cup to Po'owahe again. We passed the cup three times, the usual number for visiting guests, though I had been in kava ceremonies where the cup had been passed as many as twelve times. With that many drinks, your entire body is numb and completely relaxed. The only problem is the need to pee after so much liquid being poured into your body, but you don't leave a kava ceremony until it's over.

Po'owahe was restricted to the fale designed for him during the evening meal. The captives were fed two at a time, the gags removed only long enough for them to eat. Their king was too far away for them to say or shout anything to him, and if they'd tried the guards would have killed them. Those were the orders they'd been given, and there wasn't one of them that would have passed up a chance to do so. The prisoners knew how tenuous their position was, and used caution when un-gagged and unfettered so they could eat. They were bound and gagged again when they finished.

Torches were set up around the king's fale, which he would share with Sina and one of the guards. That way we always had someone we trusted near him through the night, and with the torches, we had visual contact. Extra torches were readied to replace those that burned out. Po'owahe complained that he wouldn't be able to sleep with the torches so close but gave in when we told him the alternative was to spend the night confined like the captives.

The fleet of enemy boats was still beyond the reef in the morning, which surprised us. Po'owahe was dragged to the shore and reminded that he and the captives would remain with us until those boats had been gone from sight for two days. Gone from sight meant they couldn't be seen from our cliff-top lookout or from a surveillance boat that would follow them for half a day. If they failed to leave, the king, along with his men, would pay the price for his treachery.

His response was to stand at the beach and wave his arms expansively, signaling to those on the boats. That seemed to be a prearranged message, and within fifteen minutes the fleet had turned toward open water and was slowly paddling away and raising their sails. About an hour later, two more of our boats came in, filled with the family of another village up the coast. They had obviously been waiting until the enemy boats left but had stayed close enough to be able to help in case of trouble.

That freed up the workload somewhat for those already at Falelima. The new arrivals, having been seen by Po'owahe, didn't need to remain hidden, so they could help with the cooking, food gathering, and guard duties. The village wasn't so empty after that, which was comforting. Po'owahe had no idea how many others were still in hiding, only that the new arrivals added to whatever strength we already had.

Later that day another family arrived in their boats. The village was full again, and it was nice to be among the bustle and activity of a thriving community once more. Po'owahe had to concede our numbers with so many more arriving, and we made sure he understood that more were on the way. Many more.

One of the families agreed to take a boat out and monitor the departing fleet of enemy boats. They were to go out a half day, make sure the enemy was leaving as promised, and remain there for two days after they were out of sight before returning. If the boats turned back at any time, the surveillance boat was to return with speed but not to engage the enemy. Po'owahe had nothing to say as our boat took to sea.

The enemy boats had smaller sails than ours because they were made with reeds and were therefore lighter. Ours were made of wood, which sat deeper in the water and needed more power to move them. Still, the enemy's sails, being dark in color, would be visible over a long stretch of ocean—three or four miles, at least, maybe more.

While we waited for the promised two days to pass, some of our people went to work gathering the food we'd promised Po'owahe and filled water skins. With all of that loaded on his boat, along with the captives, he wasn't going to make very good time. It would take him a long time to catch up with the rest of his fleet, assuming they were maintaining their southward course. They might wait for him somewhere out there, of course, and as long as they were beyond our sight, there was nothing we could do about that.

Late in the day, another family arrived, and the next morning still another. The village was as crowded as when they'd all been there. Po'owahe wouldn't know one family from another, or notice if those who'd been hiding returned to the village, so I sent a quiet message allowing them to return discreetly, a few at a time. The king would think we still had an army watching from the jungle, and another right in front of him in the village. He would have every reason to believe we were a lot stronger than what he could see.

The next day another village arrived with their two boats, and Po'owahe seemed overwhelmed by the numbers of people in the village that seemed to keep growing while he watched. He sensed the hatred The People felt for him over the loss of their loved ones and rightfully feared for his life. I assured him that he was safe as long as his people did as they'd been told, that Sina and I would protect him from anyone who might want to exact vengeance for what his people had done. Even so, he kept to himself in the fale and shook with fear each time someone came near.

It took three days before those who'd been out at sea monitoring the departure of the enemy fleet returned to report they hadn't been seen for a full two days. For the next couple of hours, we stocked up Po'owahe's boat with supplies, but since it was nearly dark, we waited to load him and his men until morning. At first light the next day, we helped his men, still gagged, onto the boat, followed by their king. They pulled away from the shore, paddled by those who'd originally brought him. Once they were beyond the reef, they raised their sails and started the long journey back to their own islands far to the south.

One of our boats, manned by enough to paddle and an eight-man security detail, escorted the king's ship until it was out of sight from land. A few hours later, it returned to report that the king had removed the gags, screamed at his men angrily, and killed one, whom he threw overboard. We were thrilled to see the last of them as they sailed away.

For another week, we kept all the villagers at Falelima in case the enemy returned, but despite having boats way out in the ocean to watch for them, they didn't appear to be coming back. We had hopefully won the peace The People so richly deserved.

Chapter 26

Before we could notify the crazy little buggers that we were done and ready to go back to our own times, The People from all the villages insisted on a celebration. As a result, another week was spent feasting and celebrating, with kava, speeches, and dances. Little Sina got caught up in the festivities and learned to do a pretty good imitation of some of the dances. She had become, as far as she was concerned, one of The People, and they accepted her as such. After all, she'd gone to war with them, killed an enemy, protected the helpless women and children, and worked alongside them in the building of the village. And her father was a native of The People.

Apu felt right at home among them. These were his people, even if they were not of his time. Still, they were closer to his way of life than anything I could offer in my time, which worried me. We were eating our breakfast when I popped the question that had been on both our minds.

"I know how much you love these people—your people—and have enjoyed being with them. If you want to stay here, I'll understand."

He didn't respond quickly. It was something he needed to think through, and my heart was breaking as I contemplated the fact that I might lose him. He was the love of my life, but his life was meant for another time. My world would never be his world. He had accepted my way of life, my customs, and society, but only out of love for me. But there was another love in his heart, a love of his people and their ways. If I truly loved him, and I did, how could I keep him from that other love?

"These are my people," he finally began with a slight tremble, "and although it is not the time I was born to, their ways are much like those of my time. I miss those ways. I miss the islands, the jungle, the foods, fiafiagas, and dances. It is a good life."

"Then you should stay with your people, Apu. You should live the life you love. Why should a man who was meant to live in the tropics have to shovel snow and eat with a fork? No, that is not you, Apu. Stay with your people. I will miss you, but I want you to be happy."

"Yes," he said, shaking his head sadly, "I should stay with my people."

My heart sank. This was it, the loss of the only man I would ever love. The man who had shared so many things with me, the father of my child.

"But they are not my people," he went on. "Who are my people? You and Little Sina, that is my people. And that is who I will stay with. I will shovel snow and eat with a fork, for my heart would not allow me to do otherwise. I love you more than any island, more than any way of doing things. It is you who I will stay with, for you are my people. If you want me to stay, you must say so. Are you wishing for me to remain here when you leave?"

"Of course not," I cried, tears welling up in my eyes. "I want you with me forever, but not if you want to be elsewhere, or in another time. I want you to be happy, that's all."

"Then it will be so, for you are the fountain from which I drink to be happy. I belong with you, not in some other place or time."

"Are you sure, honey?"

"I go with you, no matter where or when. Then I am happy."

Little Sina burst in just then and saw me crying. She stopped as if hit by an enemy club. "What's wrong, mom?"

"Nothing, baby. Not a thing. Everything is perfect."

"Oh. Well, good. You just look a little, well, unhappy, that's all."

"No, I'm happy. Very happy. I feel great."

"Okay. Anyway, it seems that our mission here is finished, isn't it?"

"I think so. There might be a few minor details to take care of, but I agree. We're pretty much done."

"That's what I thought," she said, a look of confusion on her face.

"You don't seem happy about it, baby. What's wrong?"

"I'm worried about Masana, mom. I think she's fallen in love with a guy here, but she's engaged to another guy back in our time. It isn't right."

"You think she's in love?"

"Yeah. Some guy keeps hanging around her, and she lets him. Actually encourages him, and they keep running off together."

"That would be Akime, I'm guessing."

"I think that's his name, yeah."

"Okay, I'll talk to her about it, but she's got every right to find somebody else. That jerk she was engaged to wouldn't support her, wouldn't come along to defend her or help in what was supposed to be a peaceful mission, so why should she feel an obligation to him?"

"But she made a promise, mom. She's wearing his ring."

"Not here, she isn't. It couldn't cross over, so it's still in her room back in our time. He'll get it back if she's changed her mind."

"But what if she wants to stay here, in this time? I'll be losing the only sister I've ever had."

I hadn't thought about that. If she decided to stay, I'd be losing a daughter, or what I'd come to regard as a daughter. She was as close to me as any child could be, even though she was long past being a child. She could make up her own mind now, but that wouldn't ease the pain of losing her.

"I get the distinct impression you're talking about me," Masana said, sliding into the fale silently. I looked at her sheepishly, caught in the act.

"You've been eavesdropping, haven't you?" I chided.

"Guilty as charged," she chuckled, dropping cross-legged to an open spot on the floor.

"That's bad manners, girl. But since you seem to know what we were talking about, what's your thought on the subject?"

She leaned back against one of the support posts, a grim look on her face. "I'm not sure," she began. "Yes, I think I love Akime. No, check that. I know I do, in a way that's much deeper than what I felt for Richard. He stuck with me throughout the turmoil we've faced, even when it put him in danger. But my feelings for him are much deeper than just his support for me. He's the kind of man I think I've always wanted to find."

"Does that mean you want to stay with him?"

"I'm not sure. I mean, I want to be with him, but in what time? Sometimes I think I want to stay here, and sometimes I want to take him with me back to your time. I can't make up my mind. Maybe it all depends on the Little People. What would they be willing to do? I'm confused, more than anything."

I could understand that more than anyone. Apu, my husband and lover, the father of my daughter, had been born twenty-two thousand years before me, but the Little People had been forced to bring him to my time by his own insistence. That meant it was possible, but would they do it again with Akime? There were also the legal issues to consider. In my time, things like identification and insurance were a problem for a person who legally doesn't exist as far as the world is concerned. Would we have to get fake credentials for him as we'd done with Apu and Masana?

"Trust me, I know how you feel, honey. The hardest thing I ever had to do was leave Apu behind when I left his time. It was his determination that made the big-eyed creatures bring him to me, for which I'll be forever grateful. And you know what we had to go through for you and Apu in terms of legal entanglements, so I don't need to explain that. Do you love him enough to go through that with him?"

"Of course I do, but I don't know for sure that he wants me enough to leave this time in order to have me."

"There's always the option of staying here with him, though Lord knows I don't want you to."

"Yeah, I've thought about it, and I do love this place and these people, but I don't really feel like I belong here."

"Okay, there's also Richard to think about. You'd have to break it off with him, you know. Are you ready to do that?"

"He broke it off when he refused to come with me, Mom. That's over and done."

"I think you need to give him a bit of a break, sweetie. Time travel is a scary thing for most people. Frankly, most people don't even believe it's possible. They don't understand the dynamics of it or trust the outcome. I think most people would refuse the chance to actually do it, even though it's a popular theme in fiction novels."

"But he knew you'd done it. He knew Apu had come from a long time ago, just like me, so he had to know it was safe the way we were going to do it."

"Maybe. Or maybe he didn't believe what we'd told him. Maybe he thought Apu just came from Samoa, not some ancient continent. Maybe he doesn't accept the fact that I'd done it, or that you had.

Even if he did, it's still scary to think about. Give him at least that much leeway."

"No! He had his chance and blew it."

"Have you talked with Akime about going to the future with you?"

"Of course."

"And he's willing to follow you into a world he's never seen before?"

"Hell, Mom, he's been following me ever since we came here, even when it was dangerous. I think he'd follow me into a raging volcano if I asked him to."

"I've noticed how he's been hanging around you all the time. So, he loves you and you love him, that just leaves the question of when and where you spend your lives together. Or am I missing something?"

"No, that's what it comes down to, alright. Here and now, or there and then."

"Then you'll have to make that decision together. It can't be just yours to make, or his. You have to agree on it. It won't work any other way."

"And that's the problem. He'll agree to anything I suggest, so how will I know how he truly feels?"

"Don't tell him what you want from him; ask him what he wants from you. Let him know both options are agreeable to you, and ask him which is best for him."

"I'm not sure both options are agreeable to me, though. I'm not sure I want to stay here, and I'm not sure I don't want to."

"Let him tell you his preference first, then make up your mind whether you can live with it. You have to know how he feels before you can decide how you feel. If either of you feels like he or she was coerced into whatever move is made, it'll fall apart in time. Just be honest with each other. If he wants to go into the future with you, we'll deal with the Little People when the time comes. But if he doesn't want to, there's no need to bring them into it."

"Unless I decide to stay here."

"That's true. But I want you to know that we love you very much and don't want to lose you unless it's what you truly want. We won't stand in the way of your happiness, but we'd miss you more than you could ever know."

"Thanks, Mom. You too, Dad. And especially you, Little Sina. I bet you're anxious to get back to your own time so you can just be Sina again, rather than Little Sina."

"I'm used to it," my daughter chuckled. "Sina lived with us for several months, so I had to learn to deal with it. Besides, it's her name. I was named after her, so she's got first rights to it."

"Yeah, but it's still a bit cumbersome."

"Not when you respect Sina as much as I do. She's awesome."

Just then, a young teenage boy ran up to the fale and called to me from the entrance. "Salini, chiefs wish to counsel with you."

"Which chiefs?"

"All chiefs."

"Okay, tell them I'll be there in just a moment. Will Sina be there?"

"She's there now, and waiting for you."

The meeting fale had two posts of honor. Sina was seated at one, and I was directed to the other one. Most fales have only one post of honor, but this one had been built with two to accommodate both of us. A taupou was mixing a brew in the traditional kava bowl, and began to serve it as soon as I was seated. The required speeches were made with each serving, and we went through three rounds before the ceremony was finished and we got down to the topic at hand. It was Atamiso of the Kusani family who first spoke in that sing-song narrative of the respect language.

"The People have been blessed to have warrior goddesses of old to help them in this time of war. Your song will be sung forever in our fiafiagas, and in our legends you will not be forgotten."

"Thank you, great chief Atamiso," Sina replied, "but let your own people be remembered as well, for it was they who won the battles."

"They, too, will be given praise and honor," he replied.

Masotua of Amasiata spoke next in the respect language, though he was less fluent with it than Atamiso. "Will the warriors of legend remain with us, or do you go again as has been done in the past?"

"It will be with heavy hearts that we leave, but it must be so. There are deeds that need done in other times."

Masotua nodded his understanding along with all the other chiefs, but in those nods, I thought I detected a tinge of happiness, as if they were glad that we were leaving. Sina picked up on it as well.

"But let it be known among all the people, that if a time of need should come, we will return to right any wrong that has been done. Now is the time for The People to live in peace, to help each other and share in the bounties of this, the island that is now yours. Let it be so."

"It will be so," Masotua promised, but again I had the feeling that there was something he wasn't saying, nor were any of the other chiefs. I had the distinct feeling that peace among the villages might be short-lived, which meant we might be called back sooner rather than later. They all wanted to occupy Tufutafoe, the village that lived on in the legends. Were they going to make war over it, despite the fact that an agreement had been reached?

Only time would tell. At that moment it seemed that the warrior goddesses might be called into action once again, this time against the very people we had come to help.

Chapter 27

For two days after that, Masana intentionally avoided me. I'd see her somewhere in the village, but when I tried to approach, she'd turn away and disappear. I let her have her space, knowing that she had a huge decision to make, and I didn't want to put pressure on her. During that time, I didn't see Akime at all, which made me wonder if they'd broken off their relationship. I didn't know whether to hope that was so, or regret it.

Little Sina was in a slump as well. She'd found her place among The People, a place far different from any she'd find in our time. It was a special place, a feeling of belonging and being needed that she might never again feel. What did that mean for her? Was she thinking about staying behind, never returning to her true time? Had she and Masana decided to stay together?

Apu had already made the decision to return with me to my time, but would he still feel that way if our daughter decided to remain three thousand years in my past? We all had a lot of thinking to do, and I couldn't help remembering how difficult that decision was for me when it came time for me to leave Sina in her time.

There was no question about Sina's plans. She would return to her time where she had daughters waiting for her. I would have loved having her come to stay with me in my time, but that was out of the question. Nor did I blame her. Family trumps everything else, and she had a family to raise and care for, to be with through the challenges of growing up.

My decision was clear cut as well. At least, it seemed to be, but there were moments when I questioned myself over it. Yes, I wanted to go back to the life I knew, the life I was born into, but leaving this one wouldn't be easy. I loved these people, despite the issues that had confronted us, and I loved their lifestyle. It was simple and beautiful. Now that the war was over it would become simple again, and even more beautiful.

The temperature doesn't change much from day to night in this place, but without the sun burning down on you, it feels cooler at night, even if it isn't. Darkness is a time of social interaction among The People. They sit on the edge of their fales or on the sand of the malae and visit, play their unique instruments and sing until exhaustion forces them to their mats. On what I thought would be my last night in this golden time, I joined them on the malae.

Although there were others around, I sat alone with my thoughts. It was a clear night, and the stars were bright. Stars that were too far south for me to see from my apartment in Boise. The Milky Way was like a luminescent cloud running across the sky. Where I lived, the light pollution from the city would overshadow them, making them impossible to see. I would never see that sky again. A sky that even the few torches that were burning couldn't dim. It was one of many things I would miss.

A song, lifted by a few women and one man, floated over the air. A soft melody of joy, love, and friendship. It was nothing like the hard music of my time. Softer, more personal, more captivating. I would miss that too. Just days after a war had threatened to destroy these people, they sang of life and love as if no threat had ever existed. They mourned their dead, but it didn't interfere with their love of life.

The ocean's waves hissed softly as they swept onto the beach and receded back into the depths. Sitting there, I felt a sorrowful regret that I hadn't taken more time during this mission to sit and listen to the sounds of the night, to gaze at the silver lights overhead. I should have done it more, should have brought Apu and Little Sina and Masana to share those special moments with me. And Sina.

Thinking of Sina brought a catch to my throat. Would I ever see her again? Was I going to bid a final farewell to the best friend I had ever known, and lose her in the misty fog of history? She wasn't the kind of friend I could drive across town to visit, or take a vacation and fly to where she lived. She was a woman living twenty-two thousand years in my past. No train or airline could take me to her, only the rather irritable Little People could do that, and then only when a situation arose that required the services of a warrior.

I was tired of being a warrior. Tired of having to kill, of having to fight to keep from being killed. Being called a warrior was bad enough, but a warrior goddess was worse. I was no goddess, didn't believe in goddesses. Together, Sina and I had fought the odds and miraculously survived, but we were no goddesses. We were warriors, yes, and lucky to be alive after what we'd faced, but there are no goddesses. Not us, not anyone else. I wanted to take my spear, which was extremely special to me, and go back to my time.

Finally, I went to my fale and snuggled up to Apu. He woke up and we made quiet love before falling asleep. Yes, that was a golden night. A night I knew I'd cherish as long as I lived. A night like so many I'd had on my first excursion into the past. I was going to miss such nights. Nothing in my world could match a night like that. There would be special nights in my world too, but they wouldn't be like this. My last night in a world three thousand years before I was born.

It didn't turn out to be my last night, though. Masana was the one who had to contact the Little People and let them know we were ready to go, but she continued to avoid me. Several times I tried to catch her, but she always found a way to slip away. One time she actually broke into a run that took her into the jungle. Trying to catch her seemed futile, so I decided I'd just have to wait until she came to me. She was struggling with a difficult decision, and I had to let her deal with it in her own way, no matter how long it took.

I have to admit that I was worried and forlorn myself. Little Sina and Masana were both caught up in the decision of whether to stay or go back. What if they both decided to stay? I'd be saying farewell to both of them, and Sina. My world would come crumbling down around me. I would have Apu, but my girls and best friend would never exist again in my life. Would that be reason enough for me to stay with them?

There was also the issue of what would happen if any of us remained in this time. For instance, if Little Sina just disappeared from existence in my time, there would be legal ramifications. A missing person report would certainly be filed, and I would be listed as a person of interest. No explanation of time travel would satisfy investigators, so it might be possible that I would serve time for what would be considered a murder case without a body. The same might apply to Masana, since she had friends in my time who would be concerned over her disappearance as well.

There were too many things to think about, too many questions without answers. All I could do was wait, see what decisions were made, and then deal with the fallout. What if we all stayed in this time? Would someone else face legal charges for a crime that was never committed? There had been no fallout when Sina disappeared after her stay in my time, but then she hadn't established much in the way of relationships except with my family, who knew the situation. The surgeon who had repaired her arm knew her, but it seemed that he had accepted the reality of her temporary presence in his time. His staff had come to know her, but once the recovery was complete they would think of her as just another patient that needed no further services.

With so many thoughts and concerns on my mind, I waited for Masana to make up her mind. There wasn't really much else I could do. If she decided to stay, I'd deal with whatever came of it, and if she decided to take Akime with her, we'd deal with that as well. Either way, it would be a difficult situation, but we'd dealt with it before and would again. Little Sina, if she wanted to stay, would be more difficult, but I couldn't deny her that wish if it was what she really wanted. I understood, having been there myself. This was a life she could never live in my time. A good life. A more simple life than my time could ever offer. And in many ways, a more rewarding life.

Being part of a small village or group of people provides a feeling of greater importance than you get in the corporate world, where you're nothing more than a small cog in a giant wheel. Among The People, everyone is special, fills a need upon which the village relies. Each person does his or her part for the village. They have no concept or word for personal ownership. All is done for, and belongs to, the village, so each person is as important as the next whether they gather food from the forest, fish in the ocean, or carry weapons in war. There is no hierarchy of importance. Even a chief, if he's worth his salt, must carry his weight.

"What's wrong, Mom?" I hadn't heard Little Sina's approach, so her voice took me by surprise. "What makes you think something's wrong?" I asked, turning to look at her.

She gave that twist of her head and twitch of her eyebrows that said my question was a stupid one. "I know you, Mom, and something's bothering you. You're worried about Masana, aren't you?" I nodded. "Yeah, me too."

"Do you know what she's decided?"

"No," she replied with a shake of her head. "She hasn't talked to me in days. She's actually avoiding me."

"I know. Me too."

"But she's my sister, for crying out loud."

"Maybe that's why she's avoiding you. She's afraid she might hurt you. And me."

"You mean, if she decides to stay behind with Akime?"

"Yes."

"That's stupid. If she really loves the guy, what kind of sister would I be if I made a big issue over her happiness? I'd miss her, sure, but I'd never refuse to accept her decision. I want her to be happy."

"And I want you to be happy, too, honey."

She nodded slightly, then turned to me in surprise. "Wait a minute. Are you worried that I might stay behind too?"

"Should I be?"

"No! Look, these people are great, and I love the way they think and do things, but it isn't me. I'd be bored to death in no time at all. Things are too easygoing here. I need a challenge, something to work for, to aspire to."

"I'm glad," I said, and then tears welled up in my eyes and my voice cracked, so I couldn't say more. She came over and wrapped her arms around my neck, gave me a hug I desperately needed, and kissed my cheek.

"Sorry, Mom. You're stuck with me."

We laughed together, holding each other fiercely. Apu came in just then and joined us in a three-way hug that meant the world to me. I was going to keep my family, take them back with me to my time. I didn't know about Masana yet, the very special young girl who'd come to my time to survive a cataclysmic event in hers twelve thousand years ago, when she was fourteen years old. Now she was a grown woman, but I still thought of her as my daughter—a daughter I was about to lose.

Maybe she saw the family hug, maybe not. I don't know for sure what prompted her to approach us at that specific moment. All I know is that when I saw Masana coming toward us from the edge of the village, her head low and gait slow, I knew my fears were about to be realized.

“Hey, Mom, Dad, Sina,” she began, omitting the “Little” that had become so common in reference to her sister. “We’ve gotta talk. Is it okay?”

“Of course it is, honey. We’ve been waiting for this.”

“Yeah, well, things have been sort of crazy, you know. What with all the fighting, building villages, and teaching people to hunt and gather food, and...”

“We know about all that, Masana. What we don’t know about is you and Akime.”

“Okay, let’s just get right to it. I spoke with the Little People, you know, about that. I don’t know if you realize it, Mom, but they’re kind of like pricks.”

“Trust me, I know.”

“I had to argue with them, but they finally agreed to my plan. Now I just need you to agree. It’s important to me that you understand.”

“I understand more than you know, Masana. I’ve been there before, remember?”

“I guess so,” she said with an almost imperceptible nod. “Anyway, I want to do what you did. I want to take Akime with me to your time, marry him, and have a family like you did. If that’s okay with you, I mean. I know it won’t be easy. He’ll have to learn the language and get all the necessary legal stuff. I’ll need to rely on your help for that. Is it okay? Can I bring him, and count on your help to make it work?”

I grabbed her and pulled her into another hug. I hadn’t realized how much I needed that personal contact with my girls until that moment, but it was like food to a starving person. “Of course it’s okay, baby. He’ll become part of our family, just like you.”

"You don't mind? I mean, I know it'll be a hardship for you, and I really feel bad about it, but..."

"But you love him, and that's all that matters to me. I want you to be happy."

It was her turn to engage me in another hug, fierce and loving. "Thanks, Mom. Thanks for always being there for me."

"You've been there for me too, baby. It's a two-way street, a win-win situation. Never forget that," I sobbed, holding her so tight I doubt she could breathe. "So when do we leave?"

"Tomorrow morning, right after the morning meal."

"Have you told everyone else?"

"Not yet. I didn't want to say anything until I had your approval on the whole mess. I'll go tell them now."

Chapter 28

It was no surprise that breakfast was just the opening salvo in a big farewell celebration. It was the best meal since the arrival of The People on the island, with me, Sina, Masana, Little Sina, Apu, and even Akime, treated like kings. That was followed by a long round of speeches by all the chiefs present, then a rendition of a song that had been composed about the trials and triumphs we'd faced in the war against the cannibals. Dancers accompanied the song with a frenzied display of heroism and blazing victory.

The worst part, at least for me, came at the end. Masana had arranged with the Little People for us to take a boat to Tufutafoe, where we would be returned to our individual times without being watched by the villagers. Those who had originally settled in Tufutafoe were still at Falelima, so their village, which had been occupied by the cannibals for a while, was still vacant.

It was while we waded out to the boat, along with those who would paddle us, that the song of farewell began. Warrior goddesses, I assume, are supposed to be stoic and strong, never showing any weakness, physical or emotional. In my defense, however, I'm not a warrior goddess, or a goddess of anything, and certainly didn't act like one. I was bawling like a baby. This was the third time I'd had to say goodbye to life in the past, and the second time that haunting farewell floated on the air.

There's no way to describe the power of that melody. It's more feeling than music, reaching deep into the heart and pulling all its strings in unison. Sina, who was stronger than me, tried hard to cover the tear that formed at the corner of her eyes, but the rest of us were helplessly trapped in the emotion of the moment. As it had when I had to leave Sina's time, that mournful melody wafted over the forest and across the ocean, following us for miles down the coast.

When we left the boat at Tufutafoe, those who brought us gave their final farewells, bestowing on us gifts of leis and fine mats, which we weren't sure we could take with us, then took their departure. The Little People had made it clear to Masana that they wouldn't do their thing in front of the villagers, so we simply waited on the malae while the boat slowly made its way beyond the reef and up the coast until it disappeared.

While the rest of us were, at least as far as surface appearances are concerned, calm about the upcoming trip through time, Akime was terrified. There was no way he could begin to comprehend what lay ahead, for in his world no such trip was imaginable or possible. He clung to Masana for support, as if his death was inevitable. It was only his trust in Masana that kept him from running for dear life up the trail that led back to Falelima. His trust did the job, though, and he waited with us for the appearance of the big-eyed creatures that would transport us to our disparate destinations.

One of them finally appeared, gliding slowly over the sand without actually touching it, which made Akime cling even harder to Masana. To him, such a creature had to be a demon of some kind. When it was before us it nodded at Sina, which meant she would be the first to go. I grabbed her, desperately clinging to the best friend anyone could ever have, knowing that I would never see her again. She wasn't so stoic then, and we both openly shed tears over this final goodbye.

"You have been a cherished friend," I told her, "and a great warrior. I owe you more than words can say, and I will miss you more than you can know. I will hold you forever here," I said, touching the area over my heart, "and in that way, you will always be with me."

"And you with me," she replied, "in my heart and in my mind. My memories will always be warm, and I will miss you as well."

Then, the strong warrior that she was, she broke the embrace and turned away. The black-eyed creature indicated where she should stand, then pointed the now familiar device at her, and a second later she was gone. Akime saw her simply disappear in a brilliant flash of light, and gave an audible groan of terror. Masana assured him all was well, that it was the way the Little People had sent her back to her own time.

I was still bawling over the loss of Sina. I couldn't imagine losing anyone who was so close to me in that way. She wasn't dead, so I didn't need to mourn for her, but she was gone. Twenty-two thousand years gone, back to a time known only by a very few, a time and way of life archaeologists would deny ever existed. I knew better. It existed, and I was going to prove it somehow.

Apu put an arm around my shoulders and spoke softly. "She is gone, but not gone for us. It is always that she will be with us."

"Yes," I nodded with a sob. "We shared so much."

"You fought many battles together," he agreed.

"More than that," I amended. "We shared love and happy moments. We shared celebrations and victories. We shed blood and gave blood, but more than that, we shared life—the life of kinship, of joy, and sisterhood. And she gave you to me. There will never be another like her."

"It is time," Masana broke in as the little creature came before us. With the gifts of fine mats and leis on the sand before us, we stood together. Akime was holding Masana as if to protect her from some violent force, and I held Apu on my right. Little Sina, who could now be just Sina again, stood on my left, next to the spear I had never released from my hand. The "alien" creature adjusted the dial on the device, placed it on the ground before us, and took a step back.

I wondered if the device could transport all of us at once, but the answer came when a blinding light hit us, and just like that, we were on my back lawn, right where we'd started the adventure. Akime seemed amazed at being alive and shocked by the cold and snow he was suddenly standing in. He'd never seen anything like it and didn't find it all that exciting.

None of us were dressed for the cold, wearing the same clothes we'd had in the tropical heat, and no shoes. There was a slight wind that drove the cold right through us, so we gathered up the leis and fine mats that had miraculously come with us and made a beeline into my apartment. The central air heater helped to alleviate some of the shock we suffered from the abrupt climatic change we'd just gone through. Masana laughed at Akime's reaction to the weather and tried to explain snow to him, but I think he was in too much shock to comprehend what she was saying.

"It is a demon," he choked out. "How can a person live with such as this?"

"We will get proper clothing for you," she told him, "and then you will get used to it and find it enjoyable. And when we get cold, we come in where it is warm."

He'd never experienced anything even remotely like it. In the islands, you don't hear phrases like "it's a hot one, isn't it?" or "cold day today." The weather never changes more than a few degrees, day or night. Even in a hurricane, it doesn't drop more than five degrees or so. There, the weather is either rain or sun, cloudy or clear. It was going to take Akime a year or so to adapt to the weather in a temperate zone, just as it had taken Apu and Masana. I wasn't exempt, either. After a year back in Sina's time, the severe climate change when I returned took me by surprise, even though I knew what to expect.

Once again, we were a strange family, standing in the living room dressed in the clothing of an ancient culture that no longer existed, with archaic weapons clutched in our hands. My spear could be put back in the closet, coming out only on occasion when I wanted to remember special times. It would likely never be required again. Akime had his spear, a more crude version of mine, as did Masana. Apu and Little Sina, now just Sina, had not brought their weapons with them.

Thinking of my spear brought memories of Sina. We had fought so many battles, always as a team, and now it was over. I was tired of fighting, had no desire to shed blood again, but the thought of never facing danger with Sina at my side was sobering. There is a certain thrill, a quickening of the heart, and a heightening of the sensations that comes with the onset of battle. I would miss it and abhor it, both at the same time. No more fighting was good. No more Sina, on the other hand, brought on a sadness that would never go away.

Despite my aversion to fighting, I couldn't help but wonder, maybe even hope, that another crisis might come up that would require the services of the legendary warrior goddesses. Then I would be at Sina's side again, facing whatever threat came upon us. It was a slim hope, and one that defied my dislike for combat, but I felt it nonetheless. Probably, though, I would never see Sina again, never wield my spear against an enemy as long as I lived. I could live without the battles, but living without seeing Sina again was going to be a sadness that would never go away.

But there would never be a need for ancient warriors again. The cannibals, both from Sina's time and those during the re-establishment of The People on the island that would be Savai'i, had been defeated. The villages were free and self-sufficient. In time they would spread to other islands throughout the West and South Pacific to become the great Polynesian peoples who would maintain a glorious lifestyle. Then foreign aggression and tourism would destroy it, turning it into little more than a theatrical performance.

That threat, which would overcome and eventually destroy their beautiful way of life, could not be stopped by warriors with wooden spears. For the most part, that invasion would be welcomed, even invited by the inhabitants of the islands. They would want the modern devices and conveniences the foreigners would bring with them. With paying jobs, they would shop for food rather than harvesting it themselves, watch television rather than gathering on the malae. They would have cell phones and computers and do their visiting on the internet.

In hotels and stages, they would perform a make-believe version of what they had lost, pretending to be what they no longer were. The dances, foods, gardens, and even the language would be modified to the tastes of the tourists. It would be a friendly invasion. The final conquest of The People would come without combat. It would be invited and welcomed. A beautiful way of life would die, and The People would become like the rest of us, rushing to get to work on time, struggling to pay bills.

The generations of The People in my time wouldn't even know what they had lost—what they had willingly given up. They would live in enclosed houses rather than thatched fales, would cook on electric stoves rather than in a pit, and have hot and cold water in the sink and flushing toilets. In their cars, they could drive across an island that it had taken me and Sina days to cross on foot, but they would cross it in an hour. Local police would handle any crimes committed or violent confrontations. Instead of kava, alcohol would erode the old ways.

I would mourn for the loss of fa'a, "the way," but be grateful that I had been given the chance to share it with The People when it still existed, when it still meant something special. And I would welcome any request to return to any time when it was still the way of The People, for whatever reason. Even combat? Yes, maybe even combat, if that was what I was needed for. Then Sina would be there with me, to stand at my side as she'd done so many times in the ancient past.

The sky I would stare at here in my time would be drab compared to the one I'd gazed at from a sandy malae thousands of years ago. Light pollution would make most of the stars invisible. The Milky Way wouldn't be able to shine through. I wouldn't feel the gentle breezes off the ocean or smell the fragrance of wildflowers wafting across my senses. The jungle would be replaced by tended gardens, beautiful in their own way, but not the same. Yes, there were a lot of things I was going to miss.

For now, though, I was needed here and now. Masana, Akime, Little Sina, and Apu needed me to help them navigate the time we were in. The pressures on me now weren't those in which a spear was needed, but those of a mother, wife, and spiritual guide. It was a time of peace, a time of societal pressures, of character building. I would go back to teaching and doing research, being a mother and wife, paying bills, and doing my best to give the ancient past the respect and recognition it deserved.

I knew I would relish the time, if it ever came again, that the damned Little People would take me back to the time and place I'd come to love so fiercely, with Sina as my forever companion. Whether in times of peace or war, I knew I'd go again. I'd prefer peace, but there was something so invigorating about taking up arms against an enemy in close combat that, though I hated it, would always stir my blood if Sina was with me.

Could it be possible that the time would come again when the warrior goddesses were needed? Despite my disdain for battle, my hatred for the need to kill, I knew without question that I would answer the call if it ever came again.

www.ingramcontent.com/pod-product-compliance
Lightning Source LLC
LaVergne TN
LVHW052109210926
848482LV00024B/389
9798349281648